Ecoagriculture for a Sustainable Food Future

I dedicate this book to all those past and present ecofarmers who are trying produce food sustainably for the next 1000 years and more.

Ecoagriculture for a Sustainable Food Future

Nicole Y Chalmer

© Nicole Y Chalmer 2021

All rights reserved. Except under the conditions described in the *Australian Copyright Act 1968* and subsequent amendments, no part of this publication may be reproduced, stored in a retrieval system or transmitted in any form or by any means, electronic, mechanical, photocopying, recording, duplicating or otherwise, without the prior permission of the copyright owner. Contact CSIRO Publishing for all permission requests.

The author asserts their moral rights, including the right to be identified as the author.

A catalogue record for this book is available from the National Library of Australia.

ISBN: 9781486313419 (pbk)
ISBN: 9781486313426 (epdf)
ISBN: 9781486313433 (epub)

How to cite:
Chalmer NY (2021) *Ecoagriculture for a Sustainable Food Future*. CSIRO Publishing, Melbourne.

Published by:

CSIRO Publishing
36 Gardiner Road, Cayton VIC 3168
Private Bag 10, Cayton South VIC 3169
Australia

Telephone: +61 3 9545 8400
Email: publishing.sales@csiro.au
Website: www.publish.csiro.au

Front cover: (top) aerial view of cattle herd being mustered (photo by Andrew McInnes); (bottom left) handprint in soil (photo by Nicole Chalmer); (bottom right) gnamma holes north-west of Duralynia (photo by Nicole Chalmer).

Set in 9.5/13.6 Garamond
Edited by Adrienne de Kretser, Righting Writing
Cover design by Cath Pirret
Typeset by Desktop Concepts Pty Ltd
Printed by Ingram Lightning Source

CSIRO Publishing publishes and distributes scientific, technical and health science books, magazines and journals from Australia to a worldwide audience and conducts these activities autonomously from the research activities of the Commonwealth Scientific and Industrial Research Organisation (CSIRO). The views expressed in this publication are those of the author(s) and do not necessarily represent those of, and should not be attributed to, the publisher or CSIRO. The copyright owner shall not be liable for technical or other errors or omissions contained herein. The reader/user accepts all risks and responsibility for losses, damages, costs and other consequences resulting directly or indirectly from using this information.

Acknowledgement
CSIRO acknowledges the Traditional Owners of the lands that we live and work on across Australia and pays its respect to Elders past and present. CSIRO recognises that Aboriginal and Torres Strait Islander peoples have made and will continue to make extraordinary contributions to all aspects of Australian life including culture, economy and science. CSIRO is committed to reconciliation and demonstrating respect for Indigenous knowledge and science. The use of Western science in this publication should not be interpreted as diminishing the knowledge of plants, animals and environment from Indigenous ecological knowledge systems.

Foreword

In 1977, in his seminal book *The Unsettling of America*, leading agrarian thinker Wendell Berry stated: 'An agriculture cannot survive long at the expense of the natural systems that support it and that provide it with models … We can build one system only within another. We can have agriculture only within nature, and culture only within agriculture. At certain critical points these systems have to conform with one another or destroy one another.'[i]

Agriculture in both Australia and globally has now come to *the* most critical point of its entire 10 millennia-long history. Its modern forms of industrial agriculture are now widely acknowledged to be key players in destabilising many of Earth's nine planetary systems, thereby plunging our planet deeper into the Anthropocene epoch. Agriculture is also now recognised as a key factor in the rapid escalation of many modern human health epidemics and the ongoing extinction of species globally.

Nicole Chalmer's *Ecoagriculture for a Sustainable Food Future* is an important and timely book. It deserves to become a change agent in rethinking our food and farming systems and our attitudes to nature and landscapes in this country – and indeed further afield. I believe it belongs in the same canon of such writers as Francis Ratcliffe, Eric Rolls, Tim Flannery, Bill Gammage, Bruce Pascoe and their ilk.

Aside from meticulous and prodigious research across a range of disciplines, the strength of this book is Chalmer's clear-eyed, multi-faceted overview of the state of Australia's agricultural landscapes. As such, the book is a necessary synthesis that helps us more clearly see how very few modern Australian landscape managers, and indeed people more generally, understand this land. Without such an understanding, we cannot healthily and regeneratively farm, manage and care for our landscapes.

However, in delivering what is a sobering but clear diagnosis of endemic problems, Chalmer also articulates necessary solutions.

* * *

In reminding us that agricultural systems are also Social Ecological Systems (SES), one of the book's strengths is its holistic ecological view of over 65 millennia of Australian land use by humans. Concerning Australia's first Indigenous inhabitants, Chalmer explains how their SES evolved into ecologically sustainable and resilient food-producing systems, where over 250 Indigenous nations had become effective and constructive keystone influencers in those distinct environmental systems or Countries. Such SES of culturally adaptive ecofarming, as Chalmer emphasises, were reliant upon intergenerational observational based knowledge and learning.

Contrasted with this sustainability and resilience are the rapid and devastating consequences of the second human invasion: that of European settlement and then

---

i   Berry W (1977) *The Unsettling of America: Culture and Agriculture.* Sierra Club Books, San Francisco.

subsequent immigrations. Here again, humans became the keystone influencers, but misapplied land management and philosophical/cultural views have precipitated a cascade of ongoing and accelerating destruction, simplification and loss of sustainability and resilience in our multiple landscapes.

Compounding this catalogue of disasters has been largely city-based, city-driven political systems geared to economic rationalism or short-term economic and/or political agendas – both state and federal. In being disconnected from both our landscapes and basic ecological knowledge and long-term ecological thinking, this disconnected governance has become the key driver of harm.

Underpinning this state of affairs is the fact that in our modern society no rights are ascribed to nature, combined with the widespread failure to recognise that human needs are entirely dependent on nature. Further compounding this situation, as Chalmer articulates, is that neither producers nor consumers pay for the externalities – and especially those of environmental degradation, human ill-health and/or Indigenous Country dispossession.

Chalmer rightly emphasises the concept of Social Ecological Systems. This concept is based on the fact that, traditionally, human social and ecological systems are 'deeply interconnected and co-evolving across space and time scales'. In healthy SES, with a defined boundary, neither humans nor nature have pre-eminence. Chalmer expands this thinking to include the idea that non-human animals (as being socially and culturally organised) also have a SES because they too are both part of, and also profoundly influence, ecological systems. As such, and in their approach to animals and plants, the 250-plus Indigenous nations in Australia in 1788 could rightly be considered not just ecofarmers but nations that had attained the state of long-lived SES where humans were both constructive keystone species and ecosystem engineers.

Connected to this is her emphasis on the concept of *sustainability*. This describes the ability of a food production system to exist and adapt in the long term (for more than a thousand years) without damaging the ability of evolving ecosystem structure and function to support it.

The second key concept integral to Chalmer's thesis is that of *resilience*, the subject of recent developments in ecological theory. It is important because this concept has a Janus face: it can be both positive and negative. Chalmer defines *resilience* as 'the dynamic development of complex adaptive systems that interact across time and space.' The concept is key to the idea of sustainability, because a resilient SES has the capacity to adapt and withstand disturbance, or to change and transform into new states, while still remaining capable of maintaining its functionality. The Janus face aspect is that systems degraded by poor management (especially in food production and landscape function) can become resilient in their new, less functional state and thus nearly impossible to regenerate or lift to a higher level of functionality.

The evidence is mounting that this negative side of resilience is indeed the case in Australia, where many desertified areas denuded of half a metre or more of topsoil, or vast swathes of stripped vegetation, often with deep gullies or rampant dryland salinity, may almost be impossible to fully regenerate again. These severely degraded landscapes across Australia are both socially and environmentally non-sustainable and lack resilience. Yet these are also socially highly resilient or incapable of change because they are captured by ill-formed and largely dysfunctional SES where, in this case, the key destabiliser is the 'social' component of the system. This in turn constitutes a 'lock-in pathological trap', where radical innovative, regenerative change is generally prevented by the non-ecological philosophical, economic and social rigidity of our society: that is, by the dominant world view of economic rationalism and associated dependent paradigms.

Chalmer's deep research and understanding about how SES and landscapes function are aided by her use of the parallel anthropological, ecological, historical and biophysical study of one region in Australia – the Esperance Bioregion, comprising the Sand Plain and Mallee on Western Australia's south coast – but in conjunction with an Australia-wide context. In this case the 'particular' excellently amplifies the 'general' case. As she says, 'Farming in Australia is a fragile business.' Yes: for the landscapes and their function; for biodiversity and general environmental health; for ancient Indigenous cultures; and for the health of contemporary humans (Indigenous and non-Indigenous).

Besides the world views we bring to our landscapes (a key component of any SES), at the heart of this state of affairs is the nature of modern industrial food production systems. These are widely disconnected from the consumer; most of the foods produced by them are stripped of essential nutrients and phytochemicals, while many of their genetically unnatural and modified monocultural staples are increasingly laced with harmful chemicals.

However, while food systems are a large focus in this book, we should not ignore the devastating effects on the natural environment from over-grazing and native vegetation destruction in one of the driest landscapes on Earth.

Thus Chalmer describes how the Esperance Bioregion comprised the last mass-cleared, richly biodiverse region in Western Australia, continuing through until the 1980s. Acting on the principle, in Chalmer's words, of 'why change the land use when you can change the landscape', what was inculcated was a rampant development attitude that was supported by all political factions. The consequences have included the loss of unique perennial plant-based Mallee and Sandplain ecosystems and their replacement with annual crops and pastures – leading to large-scale loss of plant and animal biodiversity and a continuously spreading salinisation in significant parts of the landscape.

Concerning the first human settlement of Australia, however, Chalmer's book has no 'Pollyanna', 'noble savage' view. She recounts mounting evidence that there was deep past human ecological destruction, implicated for example in the rapid decline of megafauna. However, after millennia of adaptation, it becomes clear that Indigenous nations had evolved long-term sustainable SES and resilient food cultures.

Tragically, these systems crumbled very quickly under the onslaught of European invasion and a subsequent cascade of disruption and destruction through the remaining 19th and 20th centuries (and indeed ongoing). And this brings us to the nub of the matter: that except for a minority of regenerative farmers and pastoralists in Australia (albeit a rapidly growing segment), the ongoing practices of modern industrial agriculture and monoculture cropping (for largely export food and fibre markets) continue to degrade the sustainability and resilience of what are very complex, long co-evolved and thus generally fragile and finely balanced SES.

Thus, the key to understanding this alarming state of affairs is that, as Chalmer points out, Australians appear to have forgotten that we and our societies are integral parts of ecological systems. The industrial extractive approach has little of the stewardship approach that would enhance the long-term health of the land. Self-renewal is crucial if Australia is to evolve long-term sustainable and resilient SES. This goes to the heart of the deeper definition of regenerative agriculture: that healthy landscape management practices should enable the inherent regenerative processes of self-organisation within complex adaptive systems. These in turn should inculcate, as Chalmer articulates, evolved, tightly bound, stable and resilient systems suited to new food and fibre regimes, just as our previous Indigenous SES had attained for their food and living cultures.

The concern now in Australia is that if modern industrial practices continue in their current form, then the majority of Australia's food and fibre producing systems will be reduced to lower levels of function (or sustainability and resilience). However, I don't believe Chalmer is arguing that we return to a hunter–gatherer SES. What is argued is that we must urgently realise that one cultural approach to land use was sustainable and resilient, the other not – and that there are serious lessons to be learnt from this.

It is my hope that this book will cut through to address the growing disconnect between our landscapes and industrial consumers. There is an utmost urgency for this, because, as Chalmer points out, conserving native ecosystems and revegetating degraded ones as perennial landscapes is still not recognised as an ecological imperative for the whole of society.

Chalmer thus powerfully argues that healthy food systems are the key to sustainability and resilience. She states that 'Food is not only a complex combination of nutrients to feed the body and brain, but a social process vital to a functioning culture and landscape belonging'. As she further points out, it was the denial of access to traditional foods that was a crucial cause of the destruction of Indigenous SES.

This in turn shines the spotlight on modern industrial food systems: of foods produced at vast distances from a disconnected consumer; food monocultures produced in ecologically dead systems (both above and below ground) and thus bereft of nutrients and phytochemicals; food not from native biodiverse systems; foods that consumers haven't grown, let alone had intimate connection with its growers; food evolved and grown in either concentrated animal feedlots or designed and engineered in laboratories and ecosystems ill-suited to indigenous ecologies; foods redolent with the wrong fats, harmfully modified proteins; and foods laced with alarming levels of industrial chemicals, pharmaceuticals and the like.

As Chalmer emphasises, what people recognise as food is culturally based, and thus, in today's modern society we have rejected the opportunities to recognise a plethora of potential foods in this ancient yet newly re-settled continent. The key point here, as leading agroecological thinker Miguel Altieri has pointed out, is that healthy food systems maintain many important elements of their surrounding natural ecosystem processes.

* * *

In many ways this book can be seen as one person's 'Royal Commission' into the past landscape degradation of not just the Esperance Bioregion and Australia's bio-geophysical landscapes – what Chalmer calls 'Paradise Lost' – but their future positive potential as well.

And so, while on the surface some of the immediate conclusions of the book appear bleak, the clear principle is that if we don't honestly understand our present planetary systems, farming landscapes and our human health predicament, then no effective solutions can be generated.

Importantly, therefore, Chalmer also presents positive solutions. These emerge as both a clearer path to comprehending SES and how landscapes function, but also are illustrated by concrete examples – including in some of Australia's most fragile pastoral rangelands and semi-arid zones. These solutions involve imaginatively applied regenerative landscape management systems that restore degraded systems.

There is an old adage in life generally, and especially in landscape management and farming in particular, that to apply regenerative and healthy solutions then one needs to deeply understand the problem. As Chalmer so thoroughly reveals in what is a courageous and clear-eyed thesis, Australian agriculture and the nation in general are not only a huge distance from arriving at this recognition, they are actually in denial that there are fundamental long-term problems with our dominant land use practices. The clear conclusion? We urgently need to re-think our current agricultural systems and combine the best of both Indigenous ecoagricultural thinking with regenerative agriculture, but also with the best of modern broad-stream agriculture.

Reading Chalmer's book will enhance this shift, as it is a clear enunciation of good and bad in past and present SES in agriculture, and also essential reading in ecological literacy – the general ignorance of which is why many modern farmers can't 'read' their landscapes and so continue to inflict ongoing harm.

But this book is not just for farmers and academics. Our modern society lacks a community-wide awareness that environmental accountability involves us all. Therefore, it is high time that Australians began to comprehend the indivisible link between natural and social capital. Such an understanding can, in turn, translate into sensible food sourcing decisions.

In many ways, this is a constructively radical book, for it deeply challenges the status quo while elegantly exposing its deadly flaws in both agricultural systems and our total society.

The simple and powerful message of this book is that we need to confront the fact that Australian landscape managers haven't yet evolved healthy SES, but instead are escalating damaging practices. And herein lies the book's radical challenge. For it holds-up a mirror to industrial agriculture and the short-term destructiveness of over-grazing of our landscapes, the ongoing clearing of native vegetation, and the annual monocultural cropping and high-input agricultural practices that continues to degrade our Earth and SES. For future survival we need to urgently adopt regenerative pathways, or else perish.

For Australia, this means adopting more adaptable, sustainable and resilient SES suited to this continent. *Ecoagriculture for a Sustainable Food Future* provides us with the insight, knowledge and tools to take that positive path. In turn, this can help us avoid the looming catastrophe by making fundamental adaptive changes to both our core values and our belief systems in regard to farming practices, to food systems, and to the way we manage our economic lives.

In 1994, Wes Jackson – a leading American plant scientist, regenerative farmer and agrarian thinker – wrote: 'Since our break with nature came with agriculture, it seems fitting that the healing of culture begin with agriculture, fitting that agriculture take the lead.'[ii] The urgency of this 'lead' has never been greater. In this richly researched, deeply thoughtful book, Nicole Chalmer provides a positive road map as to how, via a new regenerative ecoagriculture, Indigenous people, landscape managers and urban food growers and consumers alike can forge a pathway to a more equitable, sustainable and resilient agriculture and food systems that can truly address our Anthropocene challenge. It is a must-read for anyone concerned about the future.

Charles Massy
Author of *The Call of the Reed Warbler*
November 2020

[ii] Jackson, W. 1994. 'Becoming Native to This Place'.

Contents

	Foreword	v
	Acknowledgements	xii
	Cultural sensitivity warning	xiv
1.	Transformations of nature and people	1
2.	The original landscapes of nature and culture	14
3.	The first consumers of nature in Australia	30
4.	How to sustain eating nature	46
5.	Healthy ecosystems, healthy food, healthy people	61
6.	Overrun by sheep: the pastoral template for colonisation	76
7.	Ending Aboriginal social ecological systems and animal landscapes	91
8.	Civilising the bush	106
9.	Why change the land use when you can change the landscape?	121
10.	Comparing pathways, past and present: the path chosen may not let you return	137
11.	Ecoagriculture for a sustainable future	152
	Appendix	170
	Index	171

Acknowledgements

I acknowledge the Aboriginal peoples who were the land managers of the Esperance bioregion, Western Australia and Australia for many thousands of years before European invasion and colonisation. In particular I would like to thank those locally who helped with discussion and gave me oral history interviews: Sonny Graham, Trevor (Kelly) Flugge, Dorothy and Warren Dimer and Annie Dabb.

I thank my CSIRO Publishing helpers: Commissioning Editor Eloise Moir-Ford for being inspired to take on this project, Development Editor Lauren Webb who started me on developing the manuscript and Development Editor Mark Hamilton who helped me to finish. Also, all the others in CSIRO Publishing, including Tracey Kudis, who have worked towards publishing this book.

I also thank Charlie Massy, who happily agreed to write the Foreword despite his busy life and running his farm in a shocking drought period.

I was encouraged by Associate Professor Dr Nancy Cushing from the University of Newcastle to take my thesis forward into a book, and I thank her for her belief that it was worthwhile to do this.

I am very grateful to those innovative lateral thinkers, the professional ecoagriculturists who granted me interviews that were so helpful in completing the last chapter. Kim Parsons, Bob Purvis, Wendy Bradshaw, David and Brad Campbell all had in common a passion for looking after their land, sometimes at economic expense to themselves and their families, especially in the early days of regeneration. I also thank those I have not directly mentioned, with whom I talked and who provided insights into new ways of thinking.

This book was inspired by my PhD thesis *Consuming Eden: An Environmental History of Food, Culture and Nature in the Esperance Bioregion*. So, I will thank my supervisor Andrea Gaynor who took me into her funded project when I hadn't been part of the academic research world for many years. I thank my secondary supervisors: Jane Mulcock for her help in being there to have face to face discussions and helpful input, and Jenny Gregory for procedurally filling in at times when Andrea was away. From the wider project, Richard Broome was very helpful in suggesting edits for the chapters about Aboriginal people.

I am grateful to David Ward, retired fire ecologist, for his insights into human and plant fire ecology. Likewise, John Pate who spent a morning explaining bio-pedogenesis to me; and his colleague William Verboom who discussed the evolutionary and geological implications of this phenomenon and its cessation since the advent of large-scale clearing in Western Australia.

I am highly appreciative of John Simons, Brendan Nicholas, Kira Tracey, David Hall and Paul Galloway from the Department of Primary Industries and Regional Development – Agriculture and Food, who all gave me valuable information, access to their personal collections of written resources and insights concerning past and present agriculture in the Esperance Bioregion and Australia, as well as insights into future developments. I am grateful to Susan Prober and Michael O'Connell who were happy to discuss their completed research project concerning Ngadju landscape management.

Local research support came from the Esperance Museum when investigating colonial and early modern agricultural systems, and I would especially like thank past Curator Wendy Plunkett and volunteer Archivists Jennifer Ford and Mary Anne Lancaster, who gave their time to find information and images related to local agricultural history. I also thank the Esperance Museum Administration for permission to use images from the museum.

I will thank Professor Alan Cooper for giving permission to use the image of Sahel and Sunda as it was thousands of years ago.

I am grateful to the many local farmers I interviewed for my thesis, whose information has been carried forward into this book. They are Edward Hannet (since deceased), David and Dale Johnson, Geoff Grewar, Paul and Pauline Bertola, Peter and Wendy Harkness and Albert Kent. Mary Hoggart provided good local botanical information. Keith Bradby spent much time explaining to me how the Great Clearing was finally halted after enormous work and input from his group and others who were so concerned about the extinction of native ecosystems. I also thank all those who allowed me to use their images throughout my thesis. I apologise to anyone I have forgotten to acknowledge.

On a personal level, I would like to acknowledge the support and patience of my husband Phil and children Joelle, Louisa, Rohan and his wife Anita and grandsons Fynn and Jude. This venture would not have been possible without their support.

Cultural sensitivity warning

Readers are warned that there may be words, descriptions and terms used in this book that are culturally sensitive, and which might not normally be used in certain public or community contexts. While this information may not reflect current understanding, it is provided by the author in a historical context.

This publication may also contain quotations, terms and annotations that reflect the historical attitude of the original author or that of the period in which the item was written, and may be considered inappropriate today.

1
Transformations of nature and people

> Food production is the largest cause of environmental change [and has always been].[1]

> The socioeconomic activities of virtually all societies have resulted in the simplification of their natural environments.[2]

The foundational pillars of any society are amazingly simple – good water and good food. Without them, no human or non-human can exist as individuals or as societies. Producing food should not be taken for granted, for the future of Australian and all human societies depends upon the long-term sustainability of their food producing systems. Societies can only persist if the essential soil foundations and ecosystem functions on farms and their surrounding environments are healthy and sustainable. Degradation of these vital resources cannot be dismissed as an externality to unlimited economic growth – a fantastical human idea when planet Earth has finite resources. Yet in Australia economic, social and political imperatives are given far greater importance than long-term sustainable food production systems. It seems that, as outlined by Tim Flannery's book *The Future Eaters*, we in Australia are still happily eating our future and leaving a trail of degraded environments behind.[3]

Human food production systems have been key drivers of the major environmental changes throughout Australia from the deep past to the present. Food production is accepted as the largest single source of environmental change and degradation on the planet, as forests are cleared, other ecosystems lost, soils are degraded and desertified, and finite resources are exploited. It is a key factor in climate change.[1] There was a time before present agriculture where people did interact more closely with landscape ecologies and culturally recognised their dependence upon a healthy biodiverse productive nature. Food meant a huge range of edibles. It was not based upon the comparatively few species of today, for which humanity has remodelled and transformed much of the world.

Farming in Australia is a fragile business. It is particularly fragile in southern Western Australian sandplain landscapes (and parallel places throughout Australia) where there are inbuilt natural constraints that must be overcome to allow the 'normal' modern farming systems based on introduced plants and animals to continue. To grow the foods that our society regards as edible and normal, landowners must fundamentally change these ancient soils with continual inputs of mineral fertilisers containing phosphorus (P), nitrogen (N) and potassium (K), lime and gypsum and numerous essential trace minerals including cobalt (Co), copper (Cu), zinc (Zn), molybdenum (Mb) and selenium (Se). With so many inputs needed for most Western Australian farmed soils and so many nationwide, there is little ability to compete in a level playing field with other industrialised farming systems in the western world. The younger more fertile soils of Europe, the Mediterranean and the Americas where most domesticated plants and animals evolved, generally have intrinsic soil properties with greater fertility and resilience to modern farming systems. These overseas farm competitors start so far in front in their ability to grow grain, meat and other produce, it is remarkable that Australian farms perform as well as they do.

Over 63% of the Australian landmass is divided into land ownership titles or leases and is under the control or ownership of corporate and individual farmers, graziers and pastoralists. Other than pastoral leases, the land ownership laws of our cultural system mostly allow landowners to do what they want on their land. But it is time to understand that these owned lands are essentially under generational stewardship for we should be passing them on to the next generation in the same or better condition than they are now. With productivist aims dominating agriculture, there is little emphasis placed on the importance of loving Country and managing land and soils in terms of ecological knowledge and respect for their long-term future. The agribusiness attitude is that a farm is a business only, and that agricultural business owners cannot afford to, are by heritage unable to, or should not bother with forming the emotional attachment to Country that is so often cited as of utter importance to Indigenous peoples. This emotive feeling needs to be felt by all those who have the responsibility of looking after so much of Australia for future generations of people, plants, animals and ecosystems. These feelings need to be reinforced and legitimised in the activity of agriculture which is not only an economic system but a lifeway connecting ecology, cultural systems and emotional feelings of belonging. Learning about the ecology of agriculture and the landscapes in which it is conducted should be mandatory to farm or manage land in Australia.

I am connected to the land we farm at Esperance, even though its seasons are less predictable, and it lacks the rainfall of Denmark and Mt Barker where we farmed before. Here in the Esperance bioregion the land was treated cruelly during the 1960s to 1980s. There were vast unruly acts of overclearing that vandalised the riparian vegetation of rivers and creeks, lakes and wetlands – land that has only become saltier and more prone to flooding since clearing. Our farm, as one of the last sandplain farms cleared, was treated a little better with a large semi-permanent lake and smaller lakes and wetlands still surrounded by their original vegetation of freshwater paperbarks and yate forest. So much of the Esperance sandplain was indiscriminately cleared that those patches remaining have been listed as critically endangered ecological communities. They represent some of the 393 critically endangered Western Australian ecological communities.[4] While mosaics of bush and trees were retained when the Mallee region was originally cleared in the 1920s, many of these were destroyed with the introduction of bulldozers in the 1960s. They have also been listed as critically endangered ecological communities.

On our farm we have aimed to maximise habitats and biodiversity within the landscape by fencing off precious bushland areas against livestock degradation. To stabilise the once poor and drifting cleared sands with perennial grasses, the soils have needed re-engineering with quality annual legumes and grasses and a clay spreading program to improve water- and nutrient-holding capacity and the ability to build soil carbon. For the kind of agriculture that is acceptable within our culture and from which we earn money, we rely mostly upon quality non-native pastures that need soils re-engineered by generous fertilising each year with phosphorus and potash- based fertilisers which not only directly feed the plants but also indirectly feed the new commensal soil biology and animal grazers. Adding fertilisers also replaces nutrients removed in the biomass of cattle, sheep, wool and crops taken from the farm and sold. The various trace elements are also absolutely essential but are applied less frequently because of their cost. Summer rainfall events allow the perennials to flourish, turning the farm a rich living green and preventing significant runoff and wind and water erosion as well as fixing atmospheric carbon in the soils as organic carbon and plant phytoliths.[5,i]

The potential for complexity and biodiversity above ground and in soils is generally greater on farms that specialise in livestock where there is potentially greater diversity of habitats in soils, mixed pastures and fenced bush mosaics. Biological simplification is often featured on large cropping properties geared towards monocultures, which may have either poor quality or no native vegetation left. The important soil biology has declined due to yearly chemical applications and long periods of heat and dryness when nothing is allowed to grow during summer to conserve soil moisture. Because larger wildlife such as kangaroos and emus have a definite impact on returns, as they graze on and flatten crops, they are rarely tolerated. Continued salinisation due to overfilling of aquifers is also an inevitable by-product in cropping systems and annual plant based grazing landscapes for, with summer rainfall, water use does not balance rainfall at critical times.

The Esperance bioregion encompasses a huge area, approximating 42 546 km^2 surrounding the town of Esperance (population about 14 500), on Western Australia's south-east coast. It includes a variety of ecosystems found within the coastal zone, sandplain, mallee and mallee woodlands going to Israelite Bay and north to Norseman across to Fraser Range and Balladonia. Though the sandplain is typically considered a Western Australian landform, similar sandplain, mallee and woodlands ecosystems are found in in south-west New South Wales, north-west Victoria and southern South Australia. This bioregion was one of the most recently settled areas in southern Western Australia. It has been acknowledged as a hotspot of farming flexibility for its adoption of novel farming technologies and the production of innovative methods that range from technological (spreading hundreds of kilograms per hectare of clay on sandy soils to increase water and nutrient holding capacity) to biological (introducing perennial plants into pasture systems to give year-round green feed for livestock). These have allowed agri-system earnings to be improved while continuing with farming as usual. This region also represents a microcosm of the farming, environmental and economic problems afflicting farming, grazing and non-farmed landscapes Australia-wide: salinisation, soil degradation, loss of habitat, loss of biodiversity and ecosystems, increasing frequency and intensity of wildfires, water scarcity, climate change, social disruptions and the general population's cultural disconnection from the land.

---

i   Phytoliths are silicon/carbon-based biogeochemical plant structures that can last for thousands of years in soils. They could play a major role in mitigating or even reversing climate change.

As humans we rarely think about a world without us. Philosopher Francis Bacon went as far as to claim:

> Man, if we look to final causes, may be regarded as the centre of the world … insomuch that if man were taken away from the world, the rest would seem to be all astray, without aim or purpose.[6]

Before humans arrived, Australia thrived as landscapes of nature. Human activities drove rapid changes that overwhelmed the sustainability and resilience of systems, as the biodiversity and ecosystem processes of nature were destroyed. After their early damaging impacts, the first human Australian societies co-adapted with the ecosystems and species left. For at least 50 000 years they had keystone roles in maintaining relatively biodiverse ecosystems bountiful with plant and animal food. So tightly bound were most Australian ecosystems to Aboriginal land management systems, that when the next human invasion arrived in the late 18th century from Great Britain and Europe there was a rapid cascade of destructive ecosystem changes and species loss as the invasion wave spread. Colonial settlement throughout Australia directly displaced Aboriginal people from the most desirable and choicest landscape areas that had the best attributes of soil types and water availability.

To properly understand how humans fit into the world, we need to accept that we are ecological beings and in doing so accept the ecological perspectives of anthropogenic landscape change. In modern Australian food production systems research, this approach seems to have been little used. The progressive replacement of highly integrated food–culture–ecological elements within Indigenous systems with colonial and then industrialised modern food systems is accepted as a progression from past primitiveness to modern advancement. There is an implicit belief that present industrialised agricultural food systems are superior to those of the Aboriginal past. This reflects an arrogant assumption that, despite being in Australia for little more than 200 years, we recent invaders know more about managing this country than those who lived here for 50 000 years or more. At the other extreme are the unrealistic conservation science beliefs that a pre-European invasion nature can be somehow functionally restored, despite lacking Aboriginal people living with and managing ecosystems for food production day after day for generations.

The future sustainability of food production desperately needs ecological wisdom from human ecological history and the deep pre-human history of animal and plant ecosystems. This does not necessarily mean re-creating exact past ecologies, because so many of the important species and management systems are gone. However, mimics of these past ecological processes with introduced animal and plant species and modified management systems could work. There are some important ecological concepts that underpin the approaches I take in this book as we proceed through time to understand Australia's human food ecology.

Social ecological systems and agro-ecosystems

Unlike other planets in the solar system, Earth is a life-giving planet, the result of a long history of interactions between biology and geophysical processes. Living organisms fundamentally transformed the planet with a massive biodiversity explosion when they started to photosynthesise using carbon dioxide and the sun's energy for food; oxygen was produced as a waste product.[7] We as *Homo sapiens*, our relatives and ancestors have been very recent newcomers to the history of life on Earth. Nevertheless, our effects on planetary ecological systems and climate have been disproportionate, starting at least 500 000 years ago, when *Homo erectus* began using fire to manipulate landscapes.[8] New evidence points to rapidly increasing human influence, including on global climate, when humans started to farm in settlements around 8000–10 000 years ago. Most researchers now agree that humans have had environmental impacts almost everywhere on the Earth's surface.[9,10] In today's overpopulated world, human–environment interactions at nature's expense have become so accepted and transforming as to put at risk the functioning of the ecosystems necessary to support not only the needs of other life forms but also humanity.

It is important to understand some key ecological concepts, such as sustainable versus unsustainable practices, when investigating the culture and ecology of food producing systems. These concepts recognise that though there are few natural untouched systems without people (or their indirect effects), social systems cannot exist without nature. Though this seems to be common sense, for a long time after its inception ecological theory and principles were developed as though nature existed free of human influences. Ecological geographers Ian Davidson-Hunt and Fikret Berkes from the Stockholm Resilience Centre, and other researchers, were among the first to show how traditionally human social and ecological systems are deeply interconnected and co-evolving across space and time scales. They are clear that this paradigm accepts 'the evolutionary or adaptive relationships between human societies and nature where humans and their societies are part of ecological systems within a defined boundary, neither people nor nature having pre-eminence'.[11] To merge these relationships, they conclude that people live in social ecological systems (SES), and I use this concept as the guideline throughout this book.[11] I take the SES concept even further by including non-human animals as socially and culturally organised and therefore having SES, because they are part of and profoundly influence ecological systems.

The ability to adapt to change and the paths taken in adaptation can influence the long-term sustainability of human SES. In using sustainability in this context throughout this book, I am referring to the ability of a food production system to exist and adapt in the long term (1000 years and more) without doing irreparable damage to the ability of co-evolving ecosystem structures and functions to support it. This is not an unachievable idea. In the past, traditional intensive and permanent Chinese food production systems lasted for at least 4000 years (until industrialisation) as food and human waste nutrients from cities were exhaustively returned to farms.[12]

Food culture anthropologist Eugene Anderson has spent his professional life studying human food culture. He reiterates that the easy availability and abundance of food choices is too often taken for granted in modern societies such as Australia. We seem to ignore the reality that food is a fundamental prerequisite for survival and contentment at both the scale of individuals and of societies at large. What people recognise as food is culturally based (and even class based within a society), meaning that potential foods in new places may not be recognised, especially by invading cultures. These influences on production and consumption of food are so important that people will take their culturally based and often environmentally inappropriate foods and associated production systems with them when they invade new lands.[13] The type of agro-ecosystems that then develops reflects this culturally acceptable food, and can have disastrous consequences for the Indigenous SES and the ecosystems they displace. Indigenous societies, including Aboriginal Australians, had far better nutrition, based on nutrient dense varieties of plant and animal foods, than do agricultural societies that sacrificed quality for bulk starch foods mostly based on cereal grains.

Ecosystems that are re-organised for human food production purposes are essentially domesticated ecosystems which channel most of the energy produced into human food through various interventions such as use of fire, plant cultivation and animal production. They can range from highly complex biodiverse systems in which a wide range of plant and animal foods are exploited (traditional Aboriginal food systems are an example) to highly simplified ecosystems growing very few types of food (modern agricultural monocultures). Agroecologist Miguel Altieri explains that food producing systems may be considered comparatively sustainable and resilient when they maintain many important elements of their surrounding 'natural' ecosystem processes, such as recycling of manures and minerals, and maintenance of hydrology, along with environmental social responsibility, and cultural and economic viability.[14]

As part of the process of transforming nature to produce food, cultures are organised in terms of social relationships including religions, gender roles and seasonal rhythms, as well as in their basic interactions with and beliefs about nature. A community that makes its living from hunting, gathering and managing numerous resources is likely to have a different set of cultural expertise from an agricultural community that relies on relatively few plant and animal species, and will differ greatly from those raised in urban communities.[15] People in the first type of community will have sophisticated and wide ranging knowledge about their environment's interrelationships and food producing systems and their role within it. Those in the second will have knowledge about how best to produce from their few species in the vastly changed local environment, but may be quite ignorant of surrounding ecological systems, the ecosystem services they provide and the effects their agricultural practices have upon those ecosystems. Those in the third, who live in an artificial human construct world, may be exceptionally ignorant about the processes of nature they rely upon to eat, drink and breathe.

There is much to be learnt from traditional food production systems that have demonstrated their sustainability by persisting for thousands of years. Such systems are based on integrated cultural, agroecological and social principles that are not subsumed by the prevailing view that nature, culture and food are only valuable in monetary terms, as part of economic resources for a market economy. British Professor of Agro-ecology Jules Pretty has compared and contrasted traditional agricultural societies throughout Africa and western industrialised agricultural systems, for many years.[16] Key parameters were input types and quantities, the role of non-agricultural nature, and social and cultural qualities that internalise rather than externalise ecological consequences of human activity. He concludes that industrialised agricultural systems erode natural and cultural capital. Linked factors such as 'continued population growth, rapidly changing consumption patterns and the signals of climate change are driving limited resources of food, energy, water and materials towards and beyond critical thresholds.'[16] His stance is that industrialised food systems cannot be sustainable or able to maintain long-term world food production and security unless they incorporate the vital aspects of traditional systems, to develop new ways of producing food and even organising society. A small and positive example of how this can work is described by agro-ecologist Alfonso Castro, who discussed how in Kirinyaga, Kenya, the Ndia and Gichugu Kikuyu people's traditional communal agroforestry practices sustainably maintained multi-use food producing forests within their SES for generations. Colonial invasion and the consequent disruptions caused by privatisation of land tenure and other 'innovations', such as the cash crop food economy which grows food to export rather than to feed local people, prevented the continuation of traditional communal forestry. The replacement state-run agroforestry concentrates on introduced commercial monocultures but allowed traditional food forestry to continue in the commercial forests. Though inadequate when compared to the past, this does provide a degree of food security for the rural population.[17]

Aboriginal SES in the Esperance bioregion include the Nyungar and Ngadju peoples. In an Australian context, they exemplify some of the longest continuous systems of adaptive and resilient human environmental management to produce food, with archaeological and cultural evidence going back at least 15 000–20 000 years.[18,19] Though Traditional Ecological Knowledge (TEK) is slowly being recognised in Australia as important in environmental management for conservation, there is an ongoing and persistent failure to recognise that TEK systems actually managed SES within a food production framework, rather than aligning with modern conservation's aesthetic, endangered species and biodiversity aims.[20] Acknowledging that Aboriginal SES could provide information that is usable in the development of good social and ecological relationships for food production in Australia is generally not considered among producers, agricultural researchers and policy makers; rather, they are sure that modern systems are better.

Resilience, sustainability and change

There are numerous interpretations and definitions of sustainability, depending upon the context in which it is used, but we cannot take seriously any that do not recognise that the environment, its ecosystems and social systems should be regarded as interdependent. Australian ecologists Chris Cocklin and Jacqui Dibden note, 'sustainability … is an ambiguous and contested concept.'[21] It surely is, for very often it is used in the context of a definition of sustainable development. 'Sustainable development' was defined by the United Nations World Commission on Environment and Development in 1987 (the Brundtland Commission) as 'development that meets the needs of the present without compromising the ability of future generations to meet their own needs.'[22] A wide range of nations adopted this view, including the Australian federal government, which adapted it to produce a *National Strategy for Ecologically Sustainable Development*.[23,ii]

---

ii   Rees contends that 'true sustainability requires that we recognize the reality of ecological limits to material growth'.

An unrecognised problem with 'development' is that it invariably means rapid environmental change for species and ecosystems. This can be coped with by those preadapted for rapid change, but for most it means obliteration. When bushlands are cleared for farms or suburbs, the ecosystems and habitats they contain are gone. This type of development can never be considered sustainable for the ecosystems and species involved. In such cases, the use of 'sustainable' and 'development' in the same paragraph is an oxymoron – it assumes that continuous rapid economic–biophysical growth and change in a finite world is possible without serious consequences for ecosystems.[24] An alternative view of sustainability would encompass circular systems of resource and energy use that give life the chance to continue evolving in all its self-sustaining complexity and adaptability, despite change.

Sustainability in food production is therefore a continual process rather than an end point. It includes ecological, cultural and economic dimensions that interact with agricultural systems to produce long-term systems. Ecologist John Ikerd proposes that 'a sustainable agriculture can maintain its productivity and usefulness to society indefinitely' by using 'farming systems that conserve resources, protect the environment, produce efficiently, compete commercially, and enhance the quality of life for farmers and society.'[25] Charles Holling, Fikret Berkes, Carl Folke and co-researchers propose that *resilience* is the dynamic development of complex adaptive systems that interact across time and space. It is the key to sustainability because resilient SES have the capacity to adapt and withstand disturbances or, by changing and transforming into new states of being, can maintain their functionality.[26,27] However they can also be transformed for the worse (in terms of food production and biodiversity) under bad management and become resilient in their new ugly state, and difficult or impossible to regenerate.

This concept of system resilience has been applied to the Western Australian wheatbelt by ecologists Helen Allison and Richard Hobbs, who have looked at the dynamics of that region in terms of long-term sustainability and resilience to change.[28] Though this region superficially appears to be an economically adaptive system, having survived several cycles of economic fluctuations, this view fails to account for growing environmental problems such as topsoil loss and deeper physical and biological degradation; environmental pollution with cropping chemicals; accelerating and spreading hydrological imbalance and salinity; loss of biodiversity; and social factors such as population decline in country towns. Such an array of issues signifies that the wheatbelt SES is not sustainable in the long term. It is in a 'lock-in pathological trap' that is exceedingly difficult to address at the individual and landscape scales, given the economic and social rigidity of our society. The Esperance bioregion landscape was one of the last areas of Western Australia to be transformed from a landscape dominated by the perennial vegetation of trees and native bush into a virtually treeless landscape devoted to large-scale prairie-like livestock farms and annual cropping monocultures. Allison and Hobbs' conclusions are a prediction of a future trajectory that this later developed region will have to attend to, as similar environmental and social issues are already appearing.

Evidence is growing about past dramatic climatic fluctuations that would be a threat to modern food production systems that rely on climatic stability. Louise Cullen and Pauline

Grierson from the University of Western Australia reconstructed the climate history of the Western Australian south-eastern coast using tree growth rings from *Callitris* species, a small cypress conifer that grows on salt lake margins. Results showed long periods in the past where food production would not have been easy even for adaptive Aboriginal SES, as rainfall cycles included up to 30 years of aridity followed by around 15 years of good rainfall. If such conditions reappear in the present, they will be disastrous for our modern agriculture SES and the food security it represents.[29]

Rangeland researcher R.R. McAllister and others have analysed pastoralism throughout Australian rangelands.[30] They conclude that the way that most pastoralism is conducted exhibits extreme fragility and is unlikely to survive unless there are changes into new systems. The issues become especially obvious during droughts. They are rarely dealt with in an adaptive and timely manner (destocking or rotational grazing), and perennial grasses and other palatable bush vegetation are overgrazed. Whole regions become bared soil semi-deserts that then influence the local climate and hydrology, furthering the continuation of drought conditions. When rain eventually comes, additional damage is done as flooding scours away the bared topsoils and further erodes the landscape. This means that edible perennial plants can no longer become established, and salinity increases. And so, the degradation cycle continues deeper and deeper towards desertification.

In properly managed systems there are long periods of rest from grazing pressure, with rotations so that palatable perennial plants can recover and reduce the domination of annual grasses (which require consistent rainfall to germinate) and unpalatable shrubs. New evidence using a historical research approach along with field studies indicates that removing dingoes (*Canis dingo*) from pastoral lands removes control of smaller predators such as foxes and cats, and significantly contributes to the domination of unpalatable shrubs. These smaller predators prey upon the small herbivore species that eat inedible shrub seeds, and reduce their numbers. This allows an eruption of shrub seed germination at the expense of palatable perennial plants.[31] These new findings further illustrate how important it is to take a broad, whole-of-ecosystem approach to human interactions with nature. Too often, vast and harsh efforts are employed to control and eliminate individual species. This approach rarely succeeds but, if left alone with an apex predator or predators, the ecosystem sorts itself out.

Why are histories and baselines of food, culture and nature important?

Discovering the eco-environmental history of a region or place is vital, and should be mandatory before long-term plans are made on how to manage or farm it. If only short time scales are used to try and understand how the present situations have been arrived at, mistakes are inevitable. People most often believe that they live in unchanging environmental landscapes, not comprehending that their everyday activities can have profound impacts upon ecosystems, landscapes and climate in both the short and long terms. This human attribute is reflected in the social perception concept known as 'shifting baselines'. The concept was developed by fisheries ecologist Daniel Pauly to explain the acceptance of incremental lowering of catch rates and fish size in fisheries by each new generation of fisheries scientists (and fishers). He emphasised the need to recognise historical fisheries information. As he explains:

> Each new generation of fisheries scientists accepts as a baseline the stock size and species composition that occurred at the beginning of their careers and uses this to evaluate changes. When the next generation starts its career, the stocks have further declined, but it is the stocks at that time that serve as a new baseline. The result obviously is a gradual eroding of the baseline, a gradual accommodation of the creeping disappearance of resource species, and inappropriate reference points for evaluating economic losses resulting from overfishing, or for identifying targets for rehabilitation measures.[32]

This idea is widely applicable to ongoing environmental change on and off farms and pastoral stations. It cannot be recognised unless the past is historically and scientifically investigated, recorded and compared with the present.[33] Randy Olsen, also a fisheries ecologist, recognises three forms of this shifting baselines syndrome (SBS) in which each new generation:

1. lacks knowledge of how the environment used to be;
2. redefines what is 'natural', according to personal experience;
3. sets the stage for the next generation's shifting baseline.

SBS exhibits first a generational amnesia, where knowledge extinction occurs because younger generations are not aware of past biological conditions, and second, personal amnesia, where knowledge extinction occurs as individuals forget their own experiences. This allows cultural practices and institutions developed in earlier times, to incorrectly shape community expectations and beliefs even when they have contributed to present environmental problems and are inappropriate to present needs.[34]

Comparison with the past can reveal patterns that may have significance in changed and changing environmental landscapes, as well as enlightening political, institutional and social structures that have been accepted unquestioningly but that really need to be approached with new understandings of the past before future directions can be decided. For instance, in his 2016 book *Where Song Began* Tim Low provides evidence that the prehuman Australian landscapes once dominated by marsupial and reptile megafauna and birds, may have been far more fertile, productive and richer in biodiversity than has been previously accepted by researchers.[35,iii] As the concept of SBS illustrates, the incremental change in perspective on the potential productivity of nature in Australia, and the unconscious lowering of standards in respect to this productivity, would have started with the new generations of Aboriginal people born during and after the demise of megafaunal SES.[36] The outlook continues today among modern conservation scientists, agricultural scientists and food producers. They tend to assume that, in Western Australia particularly, there were always infertile and low production landscapes that needed improving before any quantity of human food could be

---

iii Low describes the overseas radiation and origin of many bird groups from Australia, where high-quality bird food nutrients were concentrated in the flora.

produced. This may well be true, if compared to geologically younger landscapes in the northern hemisphere where modern Australian agriculture originates, but in the past these landscapes were perfectly fertile and food abundant for prehuman animals and later for the Aboriginal people and species who adapted to living there.

Environmental historian Stephen Dovers relates how the cultural institutions of the first European invaders led to them (except explorer Edward John Eyre and some others, who decried this view) considering the natural landscapes of Australia as *terra nullius* – lands without title and unowned – although they were without doubt landscapes transformed and dominated by the SES of Aboriginal people and a co-evolved nature.[37] To conservationist William Lines, the European invasion was bent on a course of 'subjugation and obliteration of the bush', so that a new kind of extractive and production-based economy could be established; in 'under 200 years [the bush] … vanished before the voracious, insatiable demands of a foreign invasion.'[38] Such attitudes reflect anger for the imagined lost past, and 'solastalgia', a term created by philosopher Glenn Albrecht and colleagues to describe the depression and existential pain caused by powerlessness to prevent environmental change in homelands. These feelings are very real.[39] I have experienced them myself, when beautiful bushland areas and dairy pastures where I grew up on the outskirts of Perth, were razed and covered by vast housing estates.

Popular authors such as Lines, Geoffrey Bolton and Eric Rolls reactively berate the destructive changes wrought by European invasion yet seem unable to define what future landscapes and attitudes are acceptable. To call for a re-establishment of pre-European ecosystems is unrealistic, for it then begs the question of 'which past?' Colonial invasion impacts overlie the deep past human impacts, which again overlie the very deep past Pleistocene landscapes of nature. Realistically, it is not possible for ecosystems to remain perpetually unchanged.[40,41,42]

To produce human food from this land is unquestionably necessary and will inevitably induce future change, but how to do this in a resilient and sustainable manner is the foremost question. It is imperative to recognise that the systems being used now need rethinking, for they need to incorporate the adaptive features of complex biodiverse ecological systems. Returning to the ecoagriculture of Aboriginal SES seems unlikely, unless forced by catastrophic events, yet learning from their systems is completely necessary to long-term survival. There were early colonial adaptations that could provide useful information for present methods of food production. They exhibited some useful aspects of European traditions such as relatively local food self-sufficiency, returning organic waste to farms and deliberately adaptive livestock movements such as transhumance. These can be integrated with some Aboriginal management systems such as fire use for pasture regeneration, rotational grazing and wildfire prevention.

When Europeans first colonised Australia, they often left behind exploited and overgrazed wastelands because they had the luxury of moving on to somewhere else. Today we can no longer do this. Adaptations must be made to permanently improve and continuously regenerate what we have. Agriculture doesn't have to slavishly follow models of the past, based on rigid expectations of what farming is. We as a society and agriculturalists have become used to accommodating shifting baselines, not diligently questioning present landscape conditions as unacceptable. Comparing present and past landscape ecologies will confirm that the landscape and ecosystems of today are so different from those of the pre-European past that it is as if they were another country.

Endnotes

1. Willett W, Rockström J, Loken J, Springmann B, Lang M, Vermeulen T, Garnett S *et.al.* (2019) Food in the Anthropocene: the EAT–Lancet Commission on healthy diets from sustainable food systems. *The Lancet Commissions* **393**(10170), 447–492. doi:10.1016/S0140-6736(18)31788-4
2. Parker E (1992) Forest islands and Kayapo resource management in Amazonia: a reappraisal of the Apete. *American Anthropologist* **94**(2), 406–428. doi:10.1525/aa.1992.94.2.02a00090
3. Flannery T (1994) *The Future Eaters: An Ecological History of Australasian Lands and People.* Reed Books, Sydney.
4. Priority Ecological Communities for Western Australia Version 2817 (January 2019) Species and Communities Program, Department of Biodiversity, Conservation and Attractions.
5. Zhang X, Song Z, Hao Q, Wang Y, Ding F *et al.* (2019) Phytolith-occluded carbon storages in forest litter layers in southern China: implications for evaluation of long-term forest carbon budget. *Frontiers of Plant Science* **10**, 1–10.
6. Francis Bacon quoted in Thomas K (1991) *Man and the Natural World: Changing Attitudes in England 1500–1800.* Penguin, London.
7. Margulis L, Sagan D (1997) *Microcosmos.* University of California Press, Berkeley.
8. Ruddiman WF (2005) How did humans first alter global climate? *Scientific American* **292**, 46–53. doi:10.1038/scientificamerican0305-46
9. Russell EWB (1997) *People and Land through Time: Linking Ecology and History.* Yale University Press, New Haven.
10. Gammage W (2011) *The Biggest Estate on Earth: How Aborigines made Australia.* Allen and Unwin, Sydney.
11. Davidson-Hunt I, Berkes F (2003) Nature and society through the lens of resilience: towards a human-in-ecosystem perspective. *Navigating Social-Ecological Systems: Building Resilience for Complexity and Change.* Cambridge University Press, Cambridge.
12. King FH (1911) *Farmers of Forty Centuries; Or, Permanent Agriculture in China, Korea and Japan.* Self-published, Wisconsin.
13. Anderson EN (2005) *Everyone Eats.* New York University Press, New York.
14. Altieri MA (1995) *Agroecology: The Science of Sustainable Agriculture.* Westview Press, Boulder.
15. Worster D (1990) Transformations of the Earth: toward an agroecological perspective in history. *Journal of American History* **76**(4), 1087–1106. doi:10.2307/2936586
16. Pretty J (2011) Interdisciplinary progress in approaches to address social-ecological and ecocultural systems. *Environmental Conservation* **38**(2), 127–139. doi:10.1017/S0376892910000937
17. Castro AP (1993) Kikuyu agroforestry: an historical perspective. *Agriculture, Ecosystems & Environment* **46**, 45–54. doi:10.1016/0167-8809(93)90012-E
18. Smith M (1993) Recherché a l'Esperance: a prehistory of the Esperance region of south-western Australia. PhD thesis, University of Western Australia, Perth.
19. O'Connor M, Prober S (2010) A *Calendar of Ngadju Seasonal Knowledge: Report 1.2 to the Ngadju People.* CSIRO, Perth.
20. Muir C, Rose D, Sullivan P (2010) From the other side of the knowledge frontier: Indigenous knowledge, social–ecological relationships and new perspectives. *Rangeland Journal* **32**(3), 259–265. doi:10.1071/RJ10014
21. Cocklin C, Dibden J (2005) *Sustainability and Change in Rural Australia.* University of NSW Press, Sydney.
22. United Nations General Assembly (1987) *Report of the World Commission on Environment and Development: Our Common Future.* Annex to document A/42/427: Development and International Co-operation: Environment. Retrieved from Black A (2005) Rural communities and sustainability. In *Sustainability and Change in Rural Australia.* (Eds C Cocklin, J Dibden) pp. 20–28. UNSW Press, Sydney.

23. Commonwealth of Australia (1990, 1992, 2005) *National Strategy for Ecologically Sustainable Development*. Department of Agriculture, Water and the Environment, Canberra.
24. Rees WE (1990) The ecology of sustainable development. *The Ecologist* **20**(1), 18–23.
25. Ikerd J (1993) The need for a systems approach to sustainable agriculture. *Agriculture, Ecosystems & Environment* **46**(1–4), 147–160. doi:10.1016/0167-8809(93)90020-P
26. Folke C, Carpenter SR, Walker B, Scheffer M, Chapin T, Rockström J (2010) Resilience thinking: integrating resilience, adaptability and transformability. *Ecology and Society* **15**(4), 20. doi:10.5751/ES-03610-150420
27. Holling C, Berkes F, Folke C (1998) Science, sustainability and resource management. In *Linking Social and Ecological Systems*. (Eds F Berkes, C Folke) pp. 342–362. Cambridge University Press, Cambridge.
28. Allison HE, Hobbs RJ (2004) Resilience, adaptive capacity, and the 'lock-in trap' of the Western Australian agricultural region. *Ecology and Society* **9**(1), 3. doi:10.5751/ES-00641-090103
29. Cullen LE, Grierson PF (2009) Multi-decadal scale variability in autumn–winter rainfall in south-western Australia since 1655 AD as reconstructed from tree rings of *Callitris columellaris*. *Climate Dynamics* **33**, 433–444. doi:10.1007/s00382-008-0457-8
30. McAllister RR, Abel N, Stokes CJ, Gordon IJ (2006) Australian pastoralists in time and space: the evolution of a complex adaptive system. *Ecology and Society* **11**(2), 41. doi:10.5751/ES-01875-110241
31. Gordon CE (2015) Dingo (*Canis dingo*) extirpation and associated trophic restructuring as a mechanism influencing shrub encroachment in arid Australia. PhD thesis, University of Sydney, Sydney.
32. Pauly D (1995) Anecdotes and the shifting baseline syndrome of fisheries. *Trends in Ecology & Evolution* **10**(10), 430. doi:10.1016/S0169-5347(00)89171-5
33. Papworth SK, Rist J, Coad L, Milner-Gulland EJ (2009) Evidence for shifting baseline syndrome in conservation. *Conservation Letters* **2**, 93–100.
34. Olsen R (2002) Shifting baselines: slow-motion disaster in the sea. *Action Bioscience*, December.
35. Low T (2016) *Where Song Began*. Yale University Press, New Haven.
36. Roberts RG, Brook BW (2010) And then there were none. *Science* **327**, 420–422. doi:10.1126/science.1185517
37. Dovers S (1994) Sustainability and 'pragmatic' environmental history: a note from Australia. *Environmental History Review* **18**(3), 21–36. doi:10.2307/3984708
38. Lines WJ (1991) *Taming the Great South Land*. Allen and Unwin, Sydney.
39. Connor L, Albrecht G, Higginbotham N, Freeman S, Smith W (2004) Environmental change and human health in Upper Hunter communities of New South Wales, Australia. *EcoHealth* **1**, 47–58. doi:10.1007/s10393-004-0053-2
40. Bolton G (1992) *Spoils and Spoilers: A History of Australians Shaping their Environment*. 2nd edn. Allen and Unwin, Sydney.
41. Rolls EC (1981) *A Million Wild Acres: 200 Years of Man and an Australian Forest*. Nelson, Melbourne.
42. Rolls EC (1984) *They All Ran Wild: The Animals and Plants that Plague Australia*. Angus and Robertson, Sydney.

2
The original landscapes of nature and culture

> Some divines thought that after the day of judgement the world would be annihilated; it had been made only to accommodate humanity and would have no further use.[1]

Francis Bacon's philosophical musings about human-centric religious beliefs of his time, are unfortunately still endemic in today's world where short-term human needs are invariably put before the long-term environmental sustainability of planet Earth. In the past, the ecosystems of the Australian continent before human invasion were indifferent to humanity. The forests, bushlands, swamps, rivers, savannahs and grasslands were shaped and sustained as complex and resilient ecosystems by the interactions between animals, plants and microorganisms and the geology, geography and climate.

Life has existed on the Australian continent for at least 3.5 billion years. Tangible proof of this exists in the remains of fossilised stromatolites found in north-west Western Australia.[2] Landscapes, soils and habitats have been and are still being shaped against the background of Earth's long-term geological processes such as plate tectonics and climatic shifts as well as shorter-term processes such as human, animal and plant influences, climate, weather and fire. Because people have been modifying and manipulating Australian bioregions for so long, there are difficulties in knowing exactly what a pre-human landscape of animal and plant ecosystems would have looked like. Imagination substantiated with existing comparative examples with archaeological and palaeontological evidence can help with exploration of the deep past.

Aboriginal landscape stories

The landscape geology, geography and biological features of the Australian continent are the subject of ongoing research and explanation in western scientific terms, and are a vital part of Aboriginal cosmology. Like people everywhere, Aboriginal peoples attempted to understand how the multi-dimensional living landscape was created. Their efforts provided a living map of Country that required highly developed spatial–visual memory abilities to pass it on.[3] Anthropologists Veronica Strang and Deborah Rose believe landscape, as viewed by Aboriginal peoples, is multi-dimensional because 'it consists of people, animals, plants, Dreamings, underground, earth, soils, minerals and waters and surface water'.[4,5] Strang emphasises that this view is an alternative method of naming and organising Country. It contrasts with western culture, which classifies, surveys and 'maps' the land to form a reductive system of organising ownership and resource use.

Aboriginal peoples developed creation mythologies to name and explain Country and experience interactive relationships with the land through space and time. Dreaming stories describe how past beings created the biological and geophysical landscape and provided the Laws which determined how people were to behave and interact with Country and each other. This gave cultural meaning to the Country's prominent bio-geophysical features in terms that ensured individual and group attachment to them.[5] There is evidence from the new discipline of geomythology, that some of the Dreaming stories record memories of extinct megafauna species and past physical events, encoding valuable data about the past.[6] The Tjapwurong people of western Victoria have a legend concerning a giant emu-like bird (at least three to four times the size of the modern emu) that they call the Mihirung. Palaeontologists Peter Murray and Patricia Vickers-Rich have interpreted this as describing the extinct *Genyornis newtoni*.[7] There are stories that appear to describe the coastal inundation during the early Australian Holocene, when Tasmania and many offshore islands such as Western Australia's Recherche Archipelago were isolated from the mainland.[8]

Landscapes evolved through geology and biology

The distributions of living animal and plant species, as well as fossil evidence and radiometric dating, show that Earth is physically changing and remaking itself over immense time scales, as continental plates drift and collide over the molten magma under its surface. Rates of soil renewal through tectonic activity such as volcanoes and earthquakes, vary significantly across and within continents. Geologically, Western Australia is incredibly ancient. It has Archean eon rocks from 4400 million years ago, well before life appeared on Earth. The Proterozoic epoch rocks in the centre of Australia are younger, and even younger Phanerozoic epoch rocks underlie the eastern states. This gradation reflects Australia's inherent soil fertility, grading from the largely infertile western soils to those of greater fertility in eastern Australia.[9,10] Situated deep within the Indo-Australian plate, Australia has experienced relatively little tectonic activity for millennia. Tectonic activity is confined to the northern edge where the Indo-Australian plate collides with Asian plates, around Indonesia and New Guinea. The positions of the plates and their geological features on Earth's surface profoundly affect the configuration of ocean currents and atmospheric conditions. These in turn influence climates and therefore life.

During the Eocene epoch 50–30 million years Before Present (BP), Australia's rainfall was higher and extended far inland, with moist temperate to tropical climates.[i,9] It was a green and pleasant land of shining freshwater lakes and broad winding rivers. Sea levels were at least 300 m higher, and the southern coasts were covered by a warm shallow sea rich in sponges. With other marine animal deposits, these formed the sedimentary rocks which overlie the granitic bedrock along the southern coast. Solidified limestone hills and cliffs and underground caverns are found from the Eucla cliffs, regions north of Israelite Bay and the Nullarbor plains.[9] Sinkholes and slumps with numerous underground streams and caverns have trapped fossilised fauna, providing insights into the region's past climates, flora and fauna.[11] Recent research confirms that many Western Australian soils are not geologically derived but are made by the plants that live in them, through a process called bio-pedogenesis.[12]

---

i   Before Present years (BP) is a time scale that is now conventional in most science disciplines to specify events occurring prior to the origin of practical radiocarbon dating in the 1950s.

The Pleistocene era started 1.81 million years BP. It was a period comprising at least four Ice Age cycles – cold glaciation periods followed by warmer interglacials. The southern continental shelf was part of the mainland until sea levels rose in the early Holocene (Holocene is 11 700 BP to the present), creating many islands including Rottnest Island, the Recherche Archipelago, Kangaroo Island and Tasmania.[13] Earth today is in the interglacial period of an Ice Age.[14] Originally this was believed solely responsible for the late Pleistocene drying climate which changed the non-eucalypt, non-flammable rainforest, bushlands and grassy woodlands, to eucalypt-dominated flammable ecosystems. However, several palaeontologists and climate researchers propose that many of these dramatic vegetational changes occurred after humans arrived and coincide with the loss of large keystone herbivores which mitigated fire and influenced regional climates with their grazing patterns.[15,16]

The Esperance bioregion landscape is within the Esperance Mallee-Recherche bioregion, as per the *Interim Biogeographic Regionalisation for Australia, Version 7*. It is part of the Nyungar, Wudjarri and Ngadju Aboriginal traditional lands.[17,18] The Mallee region there, and throughout Australia, is named for the dominant mallee vegetation type. Mallee is an eastern Australian Aboriginal word (locally it was called *yeit*) describing a woody plant community dominated by distinctive small eucalypts (2–10 m) with multiple trunks arising from a common large and lumpy rootstock, called the lignotuber. This starch-storing structure allows the plant to survive and resprout after complete defoliation, usually by fire but perhaps in the past by megafauna browsers.[19] The Esperance plains and sandplains of Western Australia are typified by Kwongan vegetation of scrub and mallee heath called the *Qwowkan* plain by local Aboriginal people.[20, 21]

The total Esperance bioregion area is approximately 1.9 million hectares, including approximately 1.65 million ha of cleared farms with scattered small nature reserves often associated with swamps, lakes and granite domes. Great swathes of farmland were cleared from the 1950s until clearing bans were imposed in 1986.[22] Further east and north are reserves and unallocated crown lands, part of the Great Western Woodlands, a 16 million ha area of woodlands and heathlands interspersed with salt lakes. They are the largest relatively intact Mediterranean woodlands left in the world.[23] The area has been used for pastoralism and its now scarce native mammal fauna reflect the changes started with human invasion and colonisation. It is often described as a pristine wilderness, which omits memory of how these lands were occupied and managed by Aboriginal people for thousands of years before Europeans arrived.

Climate and climate change

Climate and weather define the state of Earth's constantly changing atmosphere within different time frames. The current state of the atmosphere determines weather in each place at a given time and local scale. Climate is the long-term state of the atmosphere in a region. It is determined by accumulated weather patterns over a long period, usually a minimum of 30 years.[24]

The climatic history of Western Australia during the Holocene period has been one of repeated fluctuations. Evidence from plant pollens indicates that a wetter climate existed in south-western Australia from 6000–5000 BP, slowly becoming drier to reach a maximum about 3200 BP, followed by another wet period, then another dry period until about 1200 BP, when it became wetter again. Now a dry period seems to be in progress.[25] Whether this is part of the natural cycle exacerbated by human bushland clearing and global warming is not really clear, though evidence is leaning strongly towards human influences.

Palaeo-climatologists Louise Cullen and Pauline Grierson have analysed the last 350 years of climate on the Western Australia south-east coast, using the tree growth rings of a native cypress (*Callitris columellaris*). Their findings show an ongoing cycle of around 15 wet years followed by 20–30 dry years that seems linked to the El Niño–Southern Oscillation index (ENSO).[26] To live comfortably through these cycles, Aboriginal social ecological systems (SES) needed to be culturally adaptive, managing for scarcity in dry periods. Periods of dryness should not be regarded as an aberration (the common viewpoint of modern agriculture) but an ongoing cyclical certainty.

Palaeo-ecologist Ian Burchard proposes that people have been impacting climate since the late Pleistocene period when 'man began to hunt', linked to large herbivore extinctions (animals greater than 1000 kg) in various world ecosystems including Australia.[27] Though extinction is not always a blitzkrieg event with rapid species disappearance, and most researchers recognise this, the loss of keystone species can trigger trophic cascades that change ecosystem structure and function, leading to further species extinctions. Wherever they have occurred, these extinctions correlate with increasing fire frequency, widespread vegetation pattern changes and worldwide climate consequences.[27] Significant evidence supports the human role in Australian megafauna extinctions and resulting cascading ecosystem changes, leading to drying impacts on Australia's climate.[11] Climate physicists Anastasia Makarieva and Viktor Gorshkov provide climatic modelling evidence that hydrological cycles and climatic conditions at both regional and continent levels in pre-historic Australia were impacted, causing a drying trend, especially inland.[16]

Western Australia's weather and climate are affected by complex influences including the oceanic/atmospheric effects of the Indian Ocean Dipole, southern annular mode, cut-off lows, sub-tropical ridge, north-west cloud bands and tropical cyclones.[28] Interwoven with these are regional climatic feedbacks owing to the vast scales of agricultural land clearing, human fires, wildfires, rangeland desertification and global warming. There is empirical evidence that at least 62% of the rainfall decline in in southern Western Australia during the last 40 years is independent of global warming, linked instead to clearing of the wheatbelt woodlands as well as extensive coastal clearing for housing developments.[29] Land degradation and desertification of the rangelands are likely also linked to rainfall decline.[30,31] Kala, Lyons and Nair's research, using the Regional Atmosphere Modelling System 6.0, shows how large-scale removal of trees and vegetation has changed the interactive dynamics of cold fronts, resulting in reduced regional rainfall.[32] Esperance farmers who live along the bushland edge describe how clouds form above the adjoining woodlands, with rain and thunderstorms following the boundary and not entering the cleared farmlands. I saw an amazing visual example when flying to Perth from Esperance in late summer along the state pest barrier fence. It was as though a transparent forcefield existed from the ground to the heavens. On the eastern green uncleared woodlands side were millions of puffy white clouds close to the ground and upwards; on the western pale beige cleared wheatbelt side there were none, except where a few clouds ventured along uncleared watercourses derived from the woodlands.

This observational link between permanent green vegetation, clouds and rainfall is supported by T.J. Lyons, whose research concludes that clearing vegetation removes its evapotranspiration warming albedo effect, significantly reducing the occurrence of convective cloud formation and therefore localised rainfall events.[33] New evidence links this further to the phenomenon of bioprecipitation, in which permanent vegetation produces raindrop nuclei.[34] Fire can also influence local and regional climate, as the gases and particulates emitted by burning vegetation and similar pollution events can significantly modify atmospheric chemical composition, reducing rainfall and altering weather and regional climate towards a drying trend.[35]

Northern hemisphere climate patterns have been imaginatively transposed onto southern Australia, described as having a Mediterranean climate with four seasons – summer, autumn, winter, spring. In contrast, south-western Australian Nyungar and Ngadju peoples recognise six main seasons along with several sub-seasons.[36] They have linked climate and weather to environmental factors including plant and animal behaviours, based on thousands of years of observations vital to their immediate well-being. These climatic interpretations were probably as precise (and maybe more) in short- and long-term weather predictions as modern meteorology. The 'weather' is an ongoing obsession for modern farmers, being so fundamentally important for agricultural production. In addition to meteorological science, local farmers have linked anecdotal observations to environmental events, such as the opinion of some that a heavy fog in late summer means there will be opening rains in 90 days.

The Western Australian climate is strongly influenced by a band of high pressure called the sub-tropical ridge. Variations in the seasons are primarily due to the movement of this ridge southwards in summer, and northwards in winter.[28] Official rainfall records for the Esperance bioregion since 1883 show climatic fluctuations with extended dry periods (droughts) occurring at intervals. These dry periods have influenced European agriculture since colonial times. A drying trend in the 1930s and 1940s that was concurrent with low prices for sheep and cattle, and the end of shepherding, led to settlers abandoning the lands north of Israelite Bay.[37] This was also the time when large-scale land clearing commenced in the region to the north, east, west and south of Salmon Gums – possibly reflecting a link between land clearing and drying of local and regional climate.[37,38]

Coupling water, plants, lands and atmosphere
Hydrology
Understanding how water is connected in an endless cycle between landscapes and atmosphere is crucial to understanding how large-scale vegetation clearing impacts the water cycle, soils and landscape health. The hydrological cycle is a highly interconnected continuous exchange of water between the atmosphere, the land, vegetation, water bodies and the oceans controlling local climatic conditions.[39] Water evaporates or is transpired from plants, stored temporarily in the atmosphere and returned to the land as rainfall. This water can be

intercepted by plants and healthy carbon-rich soil aggregates and temporarily stored before returning to the atmosphere via transpiration and evaporation. It returns to the oceans as runoff via streams and rivers, and some percolates into the groundwater. The Great Artesian Basin, which underlies 22% of Australia, holds ancient water – thousands of years old at the northern recharge points and nearly 2 million years old in the south.[40] If emptied by human activities, it would take thousands of years to replenish.

The regional climate, vegetation and soil conditions impact the potential quantity of water stored in a landscape. This can be huge even in low rainfall regions. In a 400 mm rainfall zone, the annual water deposition is about 4000 tonnes per hectare. In poorly vegetated bared soil landscapes, this is more than enough to cause topsoil-destroying runoff and oversupply to aquifers, raising salt-infused water tables. Even a small percentage change in use, such as from 94% down to 92% in large catchments, can result in excessive extra water entering groundwater systems.[41]

In resilient living landscapes, the hydrological cycle has evolved so that water coming in and water going out is balanced in a long-established sustainable equilibrium. The impacts of salts stored in the landscape are controlled by the activities of landscape-wide perennial plants using water all year round. They prevent the groundwater recharge that will bring salts into the surface soils. In drier climatic zones throughout Australia, the low rainfall means little flushing of salts in the landscape geology. These zones are especially prone to salinisation when irrigated.[42] Removal of native perennial vegetation in Australia has inevitably led to groundwater rise, salinisation and regional reductions in rainfall.

Rainfall as bioprecipitation
An extremely serious impact of removing perennial vegetation, from low ground covers to grasses, shrubs and significant trees is the impact upon local and regional climate. Biologically active landscapes that contain forests and woodlands provide a range of benefits that are at least as important as carbon sequestration. They protect and intensify the hydrological cycle associated with cooling and rainfall. By evapotranspiring, trees recharge atmospheric moisture, contributing to rainfall locally and in distant locations. They cool as they capture and redistribute energy from the sun, enhance soil water infiltration and deliver purified fresh groundwater and surface water recharge.[43] Trees and other green vegetation also produce aerosols containing microorganisms and biogenic volatile organic compounds capable of catalysing the ice crystals at near 0°C in clouds, around which raindrops form. There is increasing evidence of a bioprecipitation feedback cycle involving vegetated landscapes and the microorganisms they host. The evolutionary history of ice nucleation-active bacteria, such as *Pseudomonas syringae*, has been part of this indispensable process on geological time scales since the emergence of land plants.[43]

Salinity
Salt in Australian soils originates from two physical sources as well as recently recognised biological sources. The ancient connate salt is contained in weathered parent rock that was once the seafloor during the Eocene period. Cyclic salt originates from rain-borne sea salt

deposited in the last 10 000 years, which can range from 330 kg per hectare near the coast to 15 kg per hectare 600 km inland.[39] A third biological source is the salts brought to the soil surface by the hydraulic lift mechanism of roots of certain plants species. John Pate and William Verboom have shown that morrel trees in the mallee zone (*E. longicornis*) use bio-pedogenesis to transport calcium carbonate and salts to the soil surface via very deep taproots. They then excrete these chemicals as salts from their feeder roots into the surrounding soil to reduce understorey competition. Only salt-tolerant species such as blue bush (*Kochia spp.*), saltbush (*Atriplex* spp.) and a few grasses (*Graminae* spp.) can grow under them.[44]

Biological soils

Soils form the living skin of the Earth system, underpinning life and the productivity of food systems in terrestrial ecosystems. They are generally described as consisting of minerals, organic matter, water and air. Yet the amount of life that can live in healthy soils far outweighs life above ground, with Earth's soils estimated as containing 2500 gigatons of carbon – three times as much as in the atmosphere and four times that stored in living organisms. With arid drylands (deserts) comprising the largest terrestrial biome on Earth (about 35%), largely due to human activities, the potential to store carbon through regeneration of these lands is enormous. Large amounts of CO_2 are released into the atmosphere annually due to damaging agricultural practices, so changes to those practices will positively impact climate change.[45]

Bio-pedogenesis

Based on northern hemisphere paradigms, soils are formed and rejuvenated by geological forces such as volcanic activity, uplift of continental plates, the grinding action of glaciers and ongoing physical and biological weathering.[46] In Western Australia and other Australian regions where geological soil rejuvenation has been minimal for millions of years, the living topsoils are essentially created and maintained using biological processes – soil life, moss and lichen rock breakdown, animal composting activities and plant bio-pedogenesis. A typical soil profile is shown in Figure 2.1. It illustrates where soil life dominates and sequesters carbon and other nutrients in the topsoil, and the underlying soil horizon layers above the basement rock from which essential nutrient elements are brought to the topsoil by deep-rooted plants and animals such as earthworms and dung beetles.

Some plant species have adapted to poor soils over millions of years by developing 'niche construction' alliances with symbiotic bacteria and fungi. These species-specific microorganisms use the deep soil root-sourced elements and compounds to make new fertile topsoils for their host species, in a process called bio-pedogenesis.[47] Though Western Australian native soils are infertile for Anglo-European plants, botanist William Verboom challenges the belief of soil infertility, asserting that it is a human usage contextual concept. For the plants adapted to grow in them, these soils are perfectly fertile and bio-pedogenesis renews soils as powerfully as any geological force.[48]

Gravelly laterites found in sandy soils have been built by the proteaceous ecosystem engineering plants banksias, hakeas and grevilleas. Their specialised proteoid roots containing symbiotic bacteria are concentrated in the surface soil, forming a dense mat beneath the litter

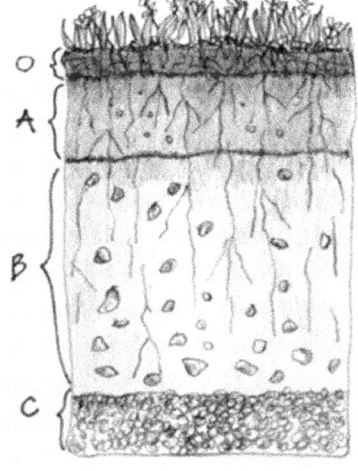

O: 0–2 cm but variable depth (humus or organic). Mostly organic matter, soil biology and plant nutrients. Perennial grass roots can go beyond 2 m deep.
A: below 0–10 cm and variable depth (topsoil). Mostly minerals from parent material with organic matter incorporated. Soil biology such as dung beetles and worms can penetrate.
B: below topsoil but variable depth (subsoil). Clay, gravel, spongolite with mineral leachates from the A horizons under Sandplain and limey nodules under Mallee. Some soil invertebrates can penetrate.
C: very deep (parent material). The deposit at Earth's surface from which the soil developed. For the Esperance Sandplain and Mallee this is ultimately granite.

Fig. 2.1: Soil horizons explained.

layer. In phosphorus-poor soils, their proteoid roots seek out traces of phosphorus and concentrate them into iron-rich nutrient-laden gravelly laterites for themselves and their seedlings.[49,50] When cleared, these gravelly soils grow the best clovers and crops, but the destruction of the original proteaceous plants has stopped this soil-forming process of bio-pedogenesis. Mallee species also engineer ecosystems, using bio-pedogenesis to create mallee soils characterised by claystone, dark grey siltstone, sandstone and brown coal (lignite), often with limestone nodules. Using specialised deep roots, they bring up chemical elements such as calcium and silicon that, with the help of symbionts, are included in clay pavement formation and nodules near the surface (in some species, limestone nodules). The formation of these clay pavements at sandplain boundaries was a slow invasion, up to 50 m in 50 years, of mallee into sandplain landscapes.[51] This soil-forming process is also destroyed by clearing of the mallee.

Mycorrhizal fungi
Over 90% of Australian native plants form symbiotic associations with mycorrhizal fungi, which receive up to 30% of plant photosynthate sugars in exchange for the hyphae actively taking up soil nutrients to share with host plants. They are keystone species, contributing to ecosystem productivity through enhancing the soil nutrient status, and are directly responsible for cycling of carbon, nitrogen and phosphorus above and below ground.[52] The mycorrhizal fungi hyphae connect at the cellular level with plant roots, greatly extending the soil area available for active water and nutrient extraction, with up to 125 000 km of hyphae in a cubic metre of healthy soil.[53] They also protect against disease organisms, exclude soil toxins and enhance plant drought resistance. They allow intra and interspecies carbon transfer and plant communication, and enhance seedling survival and growth. Native plant ecosystems throughout Australia are likely dominated by arbuscular mycorrhizal fungi of the *Gigasporaceae* family, which produce large hyphal biomass in the soil and truffle-like

fruiting bodies. Native truffles are an important food for many animals, including bettongs and bandicoots which disperse their spores. The fruiting bodies (mushrooms, truffles) of various fungal species were eaten by Aboriginal people.[54,55]

Nutrient cycling

Loss of the original keystone Pleistocene herbivores, browsers and soil diggers would have impacted nutrient recycling and fire mitigation in Australian ecosystems. Human fire use cannot do what these animals did in terms of nutrient recycling and soil building. Over millennia, fire may have impacted overall nutrient and organic matter loss as volatilisation in smoke and as leachate from soils with post fire wind and rain erosion.[56,57] The more recent widespread loss of soil digging and churning animals, such as the woylies (*Bettongia penicillata*) and mallee fowl (*Leipoa ocellata*), has further reduced the rate of decomposition and return of nutrients to native soils. These animals could move thousands of tonnes of soil per year, incorporating organic matter for decomposition and mitigating fire occurrence and intensity.[58] This suggests that 'natural' native soils of today, with often less than 1% organic matter, may actually be reflecting a relatively recent shift to low carbon infertile soils. The importance of recycling of nutrients seems to be largely unknown or disregarded in agriculture, conservation lands and the rangelands of Australia.

Modern food production in Australia requires landscape vegetation change and soil re-engineering through continuous fertiliser and trace element inputs that supply the nutrients essential for creating soils for healthy introduced plants. This is not a one-off process in south-western Australia, where yearly fertiliser applications are required, to continuously regenerate such soils. Creating these new soil ecosystems also requires species-appropriate mycorrhizal fungi and nutrient cycles that are best for plants growing in these new agro-ecosystems.[59]

Aboriginal peoples seemed well aware of soil type relationships. In some regions, it was recorded that they related food strategies to different soil types. In an 1884 publication about the Swan River Nyungar peoples, Robert Lyon described how the groups had language to describe the parallel belts of calcareous dunes, the limestone country, the low sandy dunes and the alluvium zones at the foot of the Darling scarp. The descriptions were quite independent of group boundaries and reflected a knowledge that land–plant–animal associations were distributed in predictable patterns related to soil types.[60] It is likely that other Aboriginal groups throughout Australia held similar knowledge of their Country.

Landscapes of nature, culture and fire

Ecosystem engineers and keystone species

Pre-human Australian Pleistocene landscapes are described as more fertile and wetter than they were after human arrival.[61] Ecological research has highlighted relationships between plants, animals and physical environments that ensure nutrient and energy recycling, ecosystem composition, habitats and niche structure. The large marsupial herbivores, browsers and diggers and their predators were vital as they had primary roles in creating and maintaining habitat mosaics, mitigating wildfires and providing food and shelter for themselves and co-evolved species.[11] Ecologist Clive Jones and colleagues describe such important organisms as 'ecosystem engineers', because they directly or indirectly regulate the availability of abiotic and biotic resources to other species. In so doing they modify, maintain and create habitats.[62]

Several researchers suggest that in 'a biotically mediated environment the availability of nutrients does not limit biological productivity', because the biological interactions of plants and animals play significant roles in ecosystem engineering processes. Elements such as plant and animal community composition and local biogeochemical and soil processes keep the landscape fertile.[63] Research modelling reinforces the historical and present-day importance of nutrient recycling by large herbivores. Past megafauna in Amazonia (South America) performed this extremely important ecosystem role as they transferred and distributed important nutrients such as phosphorus in dung, urine and their dead bodies.[56] This system would have been important worldwide, despite variations in the local nutrient gradients and key limiting macro and micronutrients.

The 'keystone species' concept was introduced by biologist Robert Paine in 1969. It intersects with ecosystem engineering. Like the keystones in a building, these species play such disproportionately important roles in their ecosystem's structure, function and biodiversity, that without them the ecosystem structure collapses. They range from the smallest microorganisms, fungi and invertebrates to vertebrates of all families. Their loss can cause cascading changes to ecosystem sustainability and resilience, to the point of being irreversible.[64] Bio-pedogenesis research confirms that functional niche and habitat engineering by plants is ongoing but that it can only continue if these keystone species and interacting cohorts live on. Many such relationships have been lost. For example, ecologist Greg Martin's research indicates that woylies were a keystone species. Their soil-churning behaviour composted organic matter; provided nutrients for mycorrhizal fungi and plants that improved soil structure and water-holding ability; dispersed fungal spores from the truffles they ate; and, through burying organic matter, prevented damaging wildfires in jarrah forest (*Eucalyptus marginata*).[65]

Keystone predators (or human equivalent) regulate populations of herbivores and so prevent overgrazing from damaging an ecosystem's biodiversity. Their extinction can cause a trophic cascade of change to other species. Tracey Hollings' 2013 thesis describes the role of Tasmania's apex predator, the Tasmanian devil (*Sarcophilis harrisii*), within the island's ecosystems.[ii] She confirms their essential predatory role in protecting biodiversity and maintaining ecosystem resilience. Their recent partial demise has led to cascading trophic effects and fundamental ecosystem changes in Tasmania.[66]

Animal culture

Animal cultures and their physical, biological and cultural interrelationships in ecosystems are important and complex, and have existed for far longer than human influences. Culture is an evolutionary process that speeds up transmission of skills and knowledge to others through social mechanisms. This can no longer be considered as a human-only trait.[67] The

---

ii An apex predator is the predator at the top of the food chain in an ecosystem. It profoundly influences the ecosystem's structure and functions.

time lag evident in genetic transmission of information before it becomes a population trait is inefficient compared to the speed of cultural transmission, which allows new adaptive behaviours to be learnt by many organisms in a matter of days and hours.[68] Particular forms of culture and language are related to the need for animals to pass on specific non-instinctive survival details. These include observations about their habitats and ecosystems, including predator avoidance, what foods to eat and where food and water are found.[69]

An example of large herbivore culture is found in the eastern portion of the Great Western Woodlands where the new herbivorous 'megafauna' roam, small bands of wild horses (brumbies) whose ancestors were brought to the area by European settlers during the 1870s. Until their arrival, there had been no large mammals in this landscape since the Pleistocene extinction events. The smaller mammals that survived were co-evolved to live within Aboriginal SES and many have vanished since European invasion. Palaeo-ecologists have determined that deep past ecosystems were vastly more fertile, complex and biodiverse as Pleistocene animals interacted within their ecosystems.[70] The brumbies provide a glimpse of how extinct Australian medium to large grazers could have lived in Pleistocene landscapes, as they create tracks knowingly linking landscape resources of food and water.

Ancient fire
Fire has long been an agent of landscape shaping, with a significant role in Australian ecosystems. In the prehuman past, fires started with lightning strikes during thunderstorms, their frequency related to climatic patterns linked to the pulse of glacial cycles that brought the Ice Ages, and to megafaunal grazing patterns. During interglacials, when icesheets contracted, rainfall was higher and rainforests dominated much of Australia. They were pushed out by fire-adapted sclerophyll eucalypt forest vegetation during glacial periods when icesheets expanded. Charcoal in sediments many millions of years old confirms this, as does the enormous variety of fire adaptations found in many of Australia's plant groups, including banksia, acacia and eucalypt species.[71] Features include epicormic stem shoots that resprout after fire, fruits that need fire to release seeds, and seeds that require exposure to smoke chemicals for germination.[72] Animals have also evolved fire-adaptive traits. For example, many kangaroos and wallabies flee from fire fronts then double back onto safe burnt areas; raptors hunt prey over fires and even spread fire by carrying burning branches to ignite unburnt areas. Herbivores eat fresh regrowth on burnt areas. Some species need specific fire regimes to regenerate plant communities for their special habitats.[73,74]

When grasses first appeared at around 65 million years BP, fire frequency increased, adding another factor to landscape evolution.[75] By 55 million years BP many grass species were well established, and mammalian herbivorous grazing species multiplied. Ecosystem engineering by large herbivores shaped vegetation patterns that influenced fire frequency, intensity and patterns.[76] African elephants and their tree clearing habits are a current example, for they were the major force in maintaining savannah woodlands and dependent ecosystems throughout most of Africa. Rhinoceros mitigate wildfire potential by keeping savannah grasses short and well grazed. In Australia, the large dominant herbivores were the rhino-sized diprotodon (*Diprotodon optatum*), large wombatid species and other grazing and browsing herbivores. They would have modified their landscapes, mitigated fire and perpetuated the biodiverse array of ecosystems that benefited them and coexisting species.[77]

This coevolution of grasses and grazing animals is apparent in grass adaptations. Grass tips regrow rapidly after being snipped off by grazers; grasses are high in silica; most perennial grasses need short periodic grazing to encourage regrowth and increase seed set; and many grasses are distributed by germinating in grazer dung. Grazers have evolved continuously growing powerful crowned molars to chew silica-containing grasses.[75] Fire and grazing animals benefit the continuation of grasslands – fire destroys forest trees and shrubs, allowing grasses to invade the burnt area. The grazers eat and trample any regenerating non-grazing-adapted tree and shrub species, so that the grasses maintain and extend their hold. This relationship was later exploited by Aboriginal peoples using fire to maintain grasslands for kangaroos once the megafauna grazers were gone. Since many grasses are highly flammable, their presence increases the frequency with which lightning strikes cause fire, and the cycle continues. Grazing animals are necessary to biodiversity as they reduce flammable plant biomass and create vegetation mosaics as, over the long term, unregulated wildfires homogenise vegetation ecosystems into only those that can withstand fire.[78]

When animal landscapes ruled

During the Pleistocene epoch, Ice Age periods and the evolution of grasses produced environmental conditions that stimulated the development of gigantism in animals worldwide. In Australia, this suite of fauna has been termed megafauna, though in comparison to other places fewer species were large enough (around 1000 kg) to have this title. At least 340 species of land mammal, the majority of them marsupials, inhabited Australia, as well as species of giant varanid lizards and giant land birds, the Dromornithids.[70] Palaeontological researchers have confirmed that within several thousand years of the arrival of *Homo sapiens* in the late Pleistocene (50 000–60 000+ BP), all megafauna and over 90% of the smaller co-adapted species were extinct.[76] Bone deposits found near Balladonia, in caves on the Nullarbor and at Mammoth Cave near Margaret River confirm the suite of wildlife in Western Australia as similar to those found throughout Australia in similar environments. There were giant flightless Dromornithid birds (*Genyornis*) filling the giraffe niche; 3 m high short-faced kangaroos, which walked upright, had articulated wrists and opposable thumbs, and may have filled an ape niche; rhinoceros-like diprotodon and smaller diprotodon and wombatid species that were grazers and browsers; and *Zygomaturus* species with a mobile pig-like snout, tusk incisors and powerfully clawed front paws for digging soils to get at roots and tubers.[79,80] Their influence upon soil nutrient cycling and fire mitigation would have been profound. A wide diversity of kangaroo and wallaby species were equivalents to antelope and equines. Predators included *Thylacoleo*, the marsupial lion species, thylacines (*Thylacinus* species), smaller dasyurid species and Tasmanian devils (*Sarcophilus* species). The largest predators were giant fast-running hunting lizards up to 7 m long, the *Varanus* (related to the present-day Komodo dragon).[15]

Imagine deep past Australian ecosystems before human invasion – brimming with biodiversity, vibrating and humming with life processes and very similar to recent pre-human western Madagascar.[7] Even though they had been diminished since early human arrival, the 1830 letters of biological artist John Gilbert to his employer John Gould were still able to describe bountiful Australian ecosystems and species that had over a long time period co-evolved with Aboriginal SES. These were reflected in his paintings and descriptions, written before the ecosystem-destroying impacts of European invasion had major effects.[81]

Australia today is a paradise twice lost. It will not be regained without serious societal adjustments about the nature of Australia. At present it is comprised mainly of struggling 'ecosystems' so diminished when compared to those of Africa, the Americas and the Pleistocene past. These lands still retain at least some of their important ecosystem engineers, keystone species and predators that are vital for fire mitigation, nutrient cycling and ecosystem functionality. After European invasion, the most recent fundamental changes in Australian ecosystems were triggered by enormous levels of destructive livestock overgrazing, along with tree clearing and logging. This was rapidly followed by loss of Aboriginal ecosystem management as the keystone culture, and their role as the apex hunters and predators. The persecuted dingo is now the sole apex predator (other than humans), which it wasn't in the past. Most of Australia is dominated by landscape-changing human activities, from desertifying rangelands to industrial monocultures on overcleared land. There are expanding cities and suburbs intersected with fragments of often isolated national parks and reserves, which should be more than 20 000 ha to be viable.[82] Any belief in re-creating ecosystems for the remaining endangered species seems overly optimistic. There is too much emphasis on scapegoating individual feral species, in order to justify such unsustainable and dangerous practices as dropping poison baits over thousands of square kilometres of Australian rangelands, crown lands and reserves. Instead, we Australian humans should take a real ecological keystone role with efforts focused on regenerating self-sustaining ecosystems with a variety of habitats, rather than protecting individual species. These new ecosystems should include the resilient sustainable pastoralism practised by so few, new managed 'megafauna' herbivores to mitigate fire and recycle nutrients, and predators to control herbivore populations. Trying to re-create the past without the keystone organisms begs the question 'which past?' We need to accept that change, with losses and gains, is part of the evolving web of life.[83]

Endnotes

1. Francis Bacon quoted in Thomas K (1991) *Man and the Natural World: Changing Attitudes in England 1500–1800*. Penguin, London.
2. Morrison R, Morrison M (1988) *The Voyage of the Great Southern Ark: The Four Billion Year Journey of the Australian Continent*. Ure-Smith Press, Sydney.
3. Kearins JM (1981) Visual spatial memory in Australian Aboriginal children of desert regions. *Cognitive Psychology* **13**, 434–460. doi:10.1016/0010-0285(81)90017-7
4. Rose DB (1996) *Nourishing Terrains: Australian Aboriginal Views of Landscape and Wilderness*. Australian Heritage Commission, Canberra.
5. Strang V (1997) *Uncommon Ground: Cultural Landscapes and Environmental Values*. Oxford University Press, New York.
6. Krajick K (2005) Tracking myth to geological reality. *Science* **310**(5749), 762–764. doi:10.1126/science.310.5749.762
7. Murray PF, Vickers-Rich P (2004) *Magnificent Mihirungs: The Colossal Flightless Birds of the Australian Dreamtime*. Indiana University Press, Bloomington.

8. Nunn PD, Reid NJ (2016) Aboriginal memories of inundation of the Australian coast dating from more than 7000 years ago. *Australian Geographer* **47**, 11–47. doi:10.1080/00049182.2015.1077539
9. Ward P, Brownlea D (2002) *Life and Death of Planet Earth*. Times Books, New York.
10. Johnson D (2009) *Geology of Australia*. Cambridge University Press, Melbourne.
11. Johnson CN (2009) Ecological consequences of Late Quaternary extinctions of megafauna. *Proceedings. Biological Sciences* **276**, 2509–2519. [There is vast research on the use of plant and animal fossil remains as indicators of past climate.] doi:10.1098/rspb.2008.1921
12. Verboom WH, Pate JS (2013) Exploring the biological dimensions to pedogenesis with emphasis on the ecosystems, soils and landscapes of southwestern Australia. *Geoderma* **211–212**, 154–183. doi:10.1016/j.geoderma.2012.03.030
13. Simons J (2000) *Geological History of the Esperance Region*. Department of Agriculture Western Australia, Perth.
14. Gribben J, Gribben M (2001) *Ice Age: How a Change of Climate made us Human*. Penguin, London.
15. Flannery T (1990) Pleistocene faunal loss: implications of the aftershock for Australia's past and future. *Archaeology in Oceania* **25**(2), 45–55. doi:10.1002/j.1834-4453.1990.tb00232.x
16. Makarieva AM, Gorshkov VG (2006) Biotic pump of atmospheric moisture as driver of the hydrological cycle on land. *Hydrology and Earth System Sciences Discussions* **3**(4), 2621–2673. doi:10.5194/hessd-3-2621-2006
17. Department of Sustainability, Environment, Water, Population and Communities (2012) *Interim Biogeographic Regionalisation for Australia, Version 7*. Australian Government, Canberra.
18. Palmer K (2016) Appleby Consulting Pty Ltd, consultant to native title claim through Goldfields Land and Sea Corporation. Pers. comm.
19. Retallack G (2012) Mallee model for mammal communities of the early Cenozoic and Mesozoic. *Palaeogeography, Palaeoclimatology, Palaeoecology* **342–343**, 111–129. doi:10.1016/j.palaeo.2012.05.009
20. Brooks JP (1894) Natural features of Israelite Bay. *Proceedings of the Australasian Association for the Advancement of Science, Geography Section*. Melbourne.
21. Beard JS (1990) *Plant Life of Western Australia*. Kangaroo Press, Sydney.
22. Department of Environment and Conservation (1986) *Native Vegetation Clearing Legislation in Western Australia: Environmental Protection Act 1986*.
23. Link G (2013) *The Great Western Woodlands*. <http://www.gondwanalink.org/whatsapwhere/GWWAbout.aspx>
24. Spencer C, Officer in Charge, Bureau of Meteorology office, Esperance. Interviewed by Nicole Chalmer, Esperance, 7 January 2013.
25. Seddon G (1972) *A Sense of Place*. University of Western Australia Press, Perth.
26. Cullen LE, Grierson PF (2009) Multi-decadal scale variability in autumn–winter rainfall in south-western Australia since 1655 AD as reconstructed from tree rings of *Callitris columellaris*. *Climate Dynamics* **33**, 433–444. doi:10.1007/s00382-008-0457-8
27. Burchard I (1998) Anthropogenic impact on the climate since man began to hunt. *Palaeogeography, Palaeoclimatology, Palaeoecology* **139**, 1–14. doi:10.1016/S0031-0182(97)00128-4
28. Guthrie M (2019) *Climate Drivers of the South West Land Division*. Department of Primary Industries and Regional Development, Agriculture and Food. <www.agric.wa.gov.au/climate-weather/climate-drivers-south-west-land-division>
29. Andrich MA, Imberger J (2013) The effect of land clearing on rainfall and freshwater resources in Western Australia: a multifunctional sustainability analysis. *International Journal of Sustainable Development and World Ecology* **20**(6), 549–563. doi:10.1080/13504509.2013.850752
30. Los SO, Weedon GP, North PRJ, Kaduk JD, Taylor CM, Cox PM (2006) An observation-based estimate of the strength of rainfall–vegetation interactions in the Sahel. *Geophysical Research Letters* **33**(16), L16402. doi:10.1029/2006GL027065
31. McAlpine CA, Syktus J, Ryan JG, Deo RC, McKeon GM *et al*. (2009) A continent under stress: interactions, feedbacks and risks associated with impact of modified land cover on Australia's climate. *Global Change Biology* **15**, 2206–2223. doi:10.1111/j.1365-2486.2009.01939.x
32. Kala J, Lyons TJ, Nair US (2011) Numerical simulations of the impacts of land-cover change on cold fronts in south-west Western Australia. *Boundary-Layer Meteorology* **138**, 121–138. doi:10.1007/s10546-010-9547-3
33. Lyons TJ (2002) Clouds prefer native vegetation. *Meteorology and Atmospheric Physics* **80**, 131–140. doi:10.1007/s007030200020

34. Morris CE, Conen F, Alex Huffman J, Phillips V, Pöschl U, Sands DC (2014) Bioprecipitation: a feedback cycle linking Earth history, ecosystem dynamics and land use through biological ice nucleators in the atmosphere. *Global Change Biology* **20**(2), 341–351. doi:10.1111/gcb.12447
35. Langmann B, Duncan B, Texter C, Trentmann J, van der Werf GR (2009) Vegetation fire emissions and their impact on air pollution and climate. *Atmospheric Environment* **43**(1), 107–116. doi:10.1016/j.atmosenv.2008.09.047
36. O'Connor M, Prober S (2010) A *Calendar of Ngadju Seasonal Knowledge: Report 1.2 to the Ngadju People*. CSIRO, Perth.
37. Dimer K (1989) *Elsewhere Fine*. South West Printing and Publishing, Bunbury.
38. McAlpine CA, Seabrook LM, Ryan JG, Feeney BJ, Ripple WJ, Ehrlich AH *et al.* (2007) Modelling the impact of historical land cover change on Australia's regional climate. *Geophysical Research Letters* **34**(22), L22711 doi:10.1029/2007GL031524.
39. Simons J (2000) Catchment hydrology and dryland salinity. In *Lake Warden Recovery Farm Kit*. (Ed. T Massenbauer). Department of Conservation and Land Management, Esperance.
40. Ordens CM, McIntyre N, Underschultz JR, Ransley T, Moore C (2020) Preface: advances in hydrogeologic understanding of Australia's Great Artesian Basin. *Hydrogeology Journal* **28**(1), 1–11. doi:10.1007/s10040-019-02107-8
41. Simons J, Esperance regional hydrologist. Interviewed by Nicole Chalmer, DAFWA office, Esperance, 1 February 2013–2017.
42. Short R (2000) Geological history of the Esperance region. In *Lake Warden Recovery Farm Kit*. (Ed. T Massenbauer). Department of Conservation and Land Management, Esperance.
43. Ellison D, Morris CE, Locatelli B, Sheil D, Cohen J, Murdiyarso D *et al.* (2017) Trees, forests and water: cool insights for a hot world. *Global Environmental Change* **43**(March), 51–61.
44. Pate J, Emeritus Professor of Botany, University of Western Australia. Interviewed by Nicole Chalmer, Denmark, 6 November 2014.
45. Toensmeier E (2016) *The Carbon Farming Solution: A Global Toolkit of Perennial Crops and Regenerative Agriculture Practices for Climate Change Mitigation and Food Security*. Chelsea Green Publishing, New York.
46. Jenny H (1994) *Factors of Soil Formation: A System of Quantitative Pedology*. Dover Publications, New York.
47. Verboom WH, Pate JS (2013) Exploring the biological dimensions to pedogenesis with emphasis on the ecosystems, soils and landscapes of southwestern Australia. *Geoderma* **211–212**, 154–183. doi:10.1016/j.geoderma.2012.03.030
48. Verboom WH. Phone interview by Nicole Chalmer, 27 August 2016.
49. Galloway PD, Simons J (2005) *How was our Lateritic Landscape Formed?* Department of Agriculture Western Australia, Perth.
50. Pate JS, Verboom WH, Galloway PD (2001) Occurrence of Proteaceae, laterite and related oligotrophic soils: coincidental associations or causative inter-relationships? *Australian Journal of Botany* **49**(5), 529–560. doi:10.1071/BT00086
51. Pate JS, Verboom WH (2009) Contemporary biogenic formation of clay pavements by eucalypts: further support for the phytotarium concept. *Annals of Botany* **103**(5), 673–685. doi:10.1093/aob/mcn247
52. Lambers H, Raven JA, Shaver GR, Smith SE (2008) Plant nutrient-acquisition strategies change with soil age. *Trends in Ecology & Evolution* **23**(2), 95–103. doi:10.1016/j.tree.2007.10.008
53. Wohlleben P (2015) *The Hidden Life of Trees: What They Feel, How They Communicate*. Black Inc., Melbourne.
54. Claridge AW, May TW (1994) Mycophagy among Australian mammals. *Austral Ecology* **19**(3), 251–275. doi:10.1111/j.1442-9993.1994.tb00489.x
55. Clarke PA (2011) *Aboriginal People and Their Plants*. Rosenberg Publishing, Sydney.
56. Doughty CE, Wolf A, Malhi Y (2013) The legacy of the Pleistocene megafauna extinctions on nutrient availability in Amazonia. *Nature Geoscience* **6**, 761–764. doi:10.1038/ngeo1895
57. Holt JA, Coventry RJ (1990) Nutrient cycling in Australian savannas. *Journal of Biogeography* **17**, 427–433. doi:10.2307/2845373
58. Martin G (2001) The role of small native mammals in soil building and water balance. *Stipa Native Grasses Newsletter* **16**, 4–7.
59. Van Geel M, De Beenhouwer M, Lievens B, Honnay O (2016) Crop-specific and single-species mycorrhizal inoculation is the best approach to improve crop growth in controlled environments. *Agronomy for Sustainable Development* **36**, 37. doi:10.1007/s13593-016-0373-y

60. Lyon RM (1883) A glance at the manners, and language of the Aboriginal inhabitants of Western Australia; with a short vocabulary. *Perth Gazette and Western Australian Journal*, 30 March–20 April 1883. In *Soil Guide: A Handbook for Understanding and Managing Agricultural Soils*. (Ed. G Moore) p. 18. Bulletin 4343, Natural Resource Management Services, Agriculture Western Australia, Perth.
61. Vickers-Rich P, Hewitt-Rich T (1993) *Wildlife of Gondwana*. Reed Australia, Melbourne.
62. Jones C, Lawton J, Shachak M (1994) Organisms as ecosystem engineers. *Oikos* **69**, 373–386. doi:10.2307/3545850
63. Makarieva AM, Gorshkov VG, Mackey B, Gorshkov VV (2002) How valid are the biological and ecological principles underpinning global change science? *Energy & Environment* **13**(3), 299–310. doi:10.1260/095830502320268142
64. Paine RT (1995) A conversation on refining the concept of keystone species. *Conservation Biology* **9**(4), 962–964. doi:10.1046/j.1523-1739.1995.09040962.x
65. Martin G (2003) The role of small ground-foraging mammals in topsoil health and biodiversity: implications to management and restoration. *Ecological Management & Restoration* **4**(2), 114–119. doi:10.1046/j.1442-8903.2003.00145.x
66. Hollings TA (2013) Ecological effects of disease-induced apex predator decline: the Tasmanian Devil and devil facial tumour. PhD thesis, University of Tasmania, Hobart.
67. van Schaik CP (2010) Social learning and culture in animals. In *Animal Behaviour: Evolution and Mechanisms*. Springer, Berlin.
68. Laland KN, Odling-Smee J, Feldman MW (2000) Niche construction, biological evolution, and cultural change. *Behavioral and Brain Sciences* **23**(1), 131–146. doi:10.1017/S0140525X00002417
69. Grandin T, Johnson C (2006) *Animals in Translation*. Harcourt, California.
70. Prideaux GL, Gullya GA, Couzens AMC, Ayliffec LK, Jankowskid NR, Jacobs Z et al. (2010) Timing and dynamics of Late Pleistocene mammal extinctions in south-western Australia. *Proceedings of the National Academy of Sciences of the United States of America* **107**(51), 22157–22162. doi:10.1073/pnas.1011073107
71. Abbott I, Burrows N (Eds) (2003) *Fire in Ecosystems of South-West Western Australia: Impacts and Management*. Backhuys Publishers, Leiden.
72. Flematti GR, Ghisalberti EL, Dixon KW, Trengove RD (2004) A compound from smoke that promotes seed germination. *Science* **305**, 977–977. doi:10.1126/science.1099944
73. Friend G, Wayne A (2003) Relationships between mammals and fire in south-west Western Australian ecosystems: what we know and what we need to know. In *Fire in Ecosystems of South-West Western Australia: Impacts and Management*. (Eds I Abbott, N Burrows) pp. 363–380. Backhuys Publishers, Leiden.
74. Bonta M, Gosford R, Eussen D, Ferguson N, Loveless E, Witwer M (2017) Intentional fire-spreading by 'Firehawk' raptors in northern Australia. *Journal of Ethnobiology* **37**(4), 700–718. doi:10.2993/0278-0771-37.4.700
75. Stebbins GL (1981) Coevolution of grasses and herbivores. *Annals of the Missouri Botanical Garden* **68**, 75–86. doi:10.2307/2398811
76. Rule S, Brook BW, Haberle SG, Turney CSM, Kershaw AP, Johnson CN (2012) The aftermath of megafaunal extinction: ecosystem transformation in Pleistocene Australia. *Science* **335**(6075), 1483–1486. doi:10.1126/science.1214261
77. Strömberg CAE (2011) Evolution of grasses and grassland ecosystems. *Annual Review of Earth and Planetary Sciences* **39**, 517–544. doi:10.1146/annurev-earth-040809-152402
78. Tiedeman K, District Manager, Department of Environment and Conservation, Esperance. Interviewed by Nicole Chalmer, 2011.
79. McNamara K, Murray P (2010) *Prehistoric Mammals*. Western Australian Museum, Perth.
80. Sinclair ARE (2003) The role of mammals as ecosystem landscapers. *Alces (Thunder Bay, Ont.)* **39**, 161–176.
81. Fisher CT (1985) From John Gilbert to John Gould. *Australian Zoologist* **22**(1), 5–14.
82. Main A (1984) Professor, Zoology Department, University of Western Australia. Pers. comm.
83. Wallach A (2016) Bettongs and bantengs: welcome to Australia's wild Anthropocene! *Journal of National Parks Association of NSW* **60**(1), 28–30.

3
The first consumers of nature in Australia

> We have never quite outgrown the idea, that somewhere, there are people living in perfect harmony with nature and one another, and that we might do the same if it were not for the corrupting influences of Western Culture.[1]

There is a common belief that western society First World peoples have lost the environmental connections that indigenous peoples have had for all time. It is assumed that they always lived in deep spiritual harmony and connection with their environment and in balance with nature.[2] In reality, the ancestral dispersion of *Homo sapiens* out of Africa into lands previously unoccupied by humans does not support this belief. All over the world, early invading peoples had profound destructive impacts on the new ecosystems they encountered. The first human invaders of Australia were no different. They had devastating impacts on the original Pleistocene animal and plant ecosystems of nature and culture before eventually evolving co-adapted social ecological systems (SES).[3,4]

The first peoples who invaded Sahul-Australia (the great Southland) had SES suited to Africa where humans had evolved, though as they travelled southwards new SES would have developed. Entering the Australian continent would have been like arriving in Eden – a land of maximum biodiversity, the result of eons of evolutionary and social ecological interactions between very large and small animals, plants, fungi and microorganisms and their geophysical world. All had evolved without humans, so had not developed instinctive genetic or learnt cultural adaptations to withstand human predatory behaviours.[5] Here there were truly non-human cultural landscapes, shaped and managed for eons by ecosystem engineers, keystone species and animal cultures to produce resilient, dynamic and biodiverse ecosystems, with long histories of adaptation to major perturbations such as climate and sea level changes.[6]

The entry of people and their food exploitation behaviours characterised the natural phenomenon Brian Silliman and co-researchers term a 'consumer front'. These occur when concentrated exploitation locally overwhelms the birth rate of prey. The consumers collectively move from prey-depleted areas to adjacent prey-abundant habitats along the edge of the remaining prey population. Once formed, consumer fronts move through systems as spatially propagating waves, self-reinforced via intense overexploitation which can lead to ecosystem collapses at various levels. The interwoven threads of Sahul-Australian landscapes of nature were severely disrupted and disconnected in a period of overexploitation, during which over 90% of the original marsupial and non-marsupial megafauna and their cohorts became extinct.[7] The changed ecosystems were of greatly reduced resilience and biodiversity, and extremely vulnerable to wildfires.

To survive ecosystem collapses throughout Australia, people needed to make cultural adaptations to take on the ecological roles of apex predator, keystone culture and ecosystem engineer for sustainable ecosystem services. New knowledge systems and emotional linkages

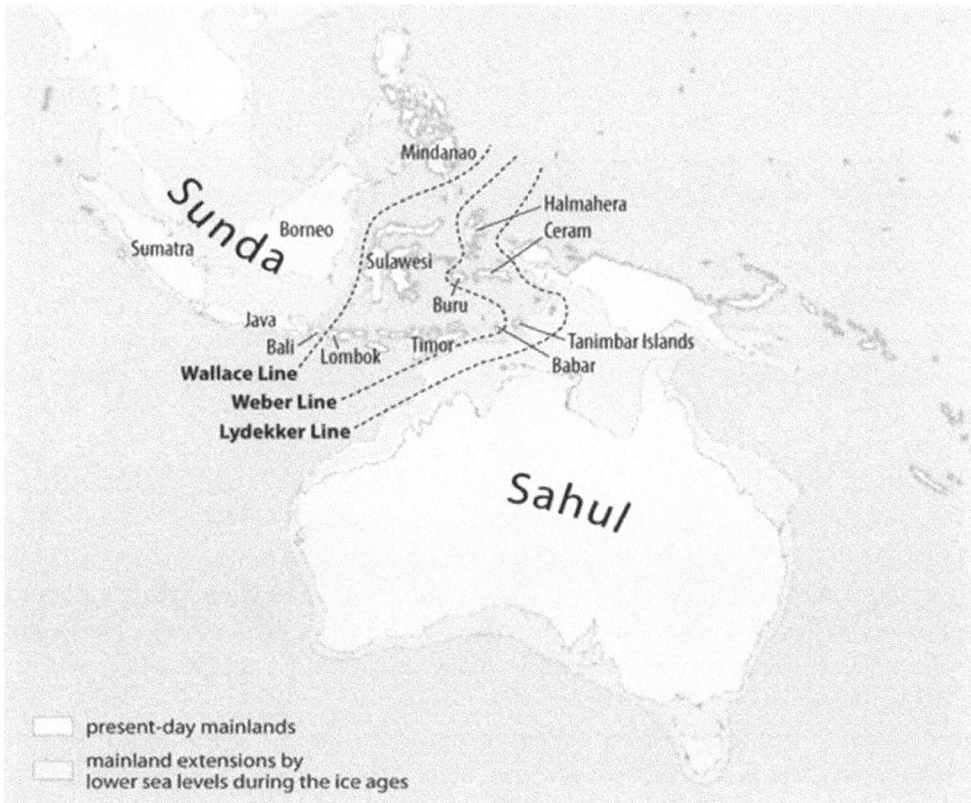

Fig. 3.1: Position of regions and appearance of Sahul, 8300–115 000 BP.[11]

were essential to produce food and best manage a country emptied of most of the original animal cultures, keystone species and ecosystem engineers who had owned and managed the ecosystems for millennia.[8] Ancestral Australians developed adaptive SES over 50 000–60 000 years or more in a dynamic process, intersected by periods of dramatic climatic oscillations and sea level changes.[9] Historical documentation in journals and diaries by early European explorers and settlers portrays a range of cultural adaptations and fire technology practices that had transformed Australian landscapes and ecosystems into enduring Aboriginal SES after the original landscapes of nature were lost.

Deep origins

From at least 115 000 to about 8300 years BP (Before Present), Australia (including Tasmania) and New Guinea formed a continuous enormous landmass called Sahul.[10] It was separated from the western landmass of Sunda by a deep ocean trough with strong currents, indicated by the Wallace Line.[i] The only mammals to breach this barrier in the last 60 000 years have

---

i   The Wallace Line was drawn in 1859 by British naturalist and co-founder of the theory of evolution by natural selection, Alfred Russel Wallace. Named by English biologist Thomas Henry Huxley, it biogeographically separates the fauna of Asia and Australia. In 1895, British naturalist Richard Lydekker further delineated the biogeographical boundary through Indonesia. Known as Lydekker's Line, that follows the Sahul Shelf edge, it separates Wallacea on the west from Australia–New Guinea on the east.

been humans, bats, rats and mice. It is possible that in the much deeper past the Denisovans (related to Neanderthals) may have also made the crossing.[11] The Wallace Line marks the boundary between Oriental and Australian faunal regions. Geographically it follows the Sunda shelf and, though no longer considered a regional boundary by many zoogeographers, it does represent a change in distribution and abundance of many vertebrate groups from fish to mammals. Lydekker's Line follows the Sahul shelf and marks the boundary of marsupial-dominated ecosystems of New Guinea and Australia to the east, as shown in Fig. 3.1.

During this time, the coastline of Sahul fluctuated as ice retreated or advanced at the poles. The closest point between the two landmasses was at Timor, considered the most likely place for people to have crossed into Sahul during the window of opportunity presented by low sea levels around 63 000–70 000 BP. Sea levels were 80–100 m below present levels, and the distance to travel by sea craft was 70–150 km.[5] Sea levels rose rapidly around 60 000 BP as polar ice melted, and the distance to reach mainland Australia increased to at least 220 km. This date of human arrival is supported by recent findings at Madjedbebe, a rock shelter in northern Australia which Chris Clarkson and co-researchers have dated at around 65 000 BP.[12] This is earlier than previously thought. The timing is supported by new dating techniques such as those used at Devil's Lair at Margaret River, which pushes back that occupation to at least 47 000–50 000 BP.[13] Earlier coastal occupation sites are likely to have been inundated when sea levels rose, so present dating may underestimate arrival times by thousands of years.[14,ii]

Alan Cooper and Christopher Stringer suggest that the first human settlers to enter Sahul may not have been modern humans at all, but Denisovans.[11] The genetic research of Irena Pugach and Anna-Sapfo Malaspinas and co-researchers strongly supports this view with evidence that Australian Aboriginal peoples, Mamanwa (a negrito group from the Philippines) and New Guinean highlanders all share DNA and skeletal structure properties that support a common origin with a divergence time around 35 000–37 000 BP.[15,16] They not only have 1–2% Neanderthal DNA, as found in other non-African humans, but also share ancestry with the Denisovan branch of humanity who were related to Neanderthals. DNA analysis shows that New Guineans, Mamanwa and Australian Aboriginal peoples have the highest percentage of Denisovan DNA (about 4–6%) of any human lineage. It may indicate that Sahul was populated by Denisovan peoples well before modern humans arrived.

Modern humans were possibly a secondary dispersal group originating from the initial dispersal of *Homo sapiens* out of Africa around 70 000 BP.[17] It may have taken as little as 3000 years for these people to reach Sahul from the Indian subcontinent, following coastal marine food resources dominated by shellfish. Shellfish are easily overexploited, and this may have caused continual coastal movement to find new unexploited areas. Middens of shellfish and other marine organisms are found at the very earliest occupation sites in coastal Australia and Timor.[18]

It is unlikely that ancestral Australians were a culturally and genetically isolated population. There is strong genetic evidence of substantial Holocene gene flow of about 11%

---

ii A detailed overview of new dating methodology pushing back the time frame of human occupation.

between a migrant Indian population and Australian Aboriginal peoples at about 4000–6000 BP. This supports the Dravidian-speaking groups as the best match for a source population.[16]

The length of time that Aboriginal peoples have occupied the Esperance bioregion is undetermined. Moia Smith's 1993 PhD research examines one of the oldest dated prehistory sites at Cheetup Hill, a large granite dome about 6 km from the coast and 65 km east of Esperance in the Cape Le Grand National Park. Her work shows that Aboriginal people were at this site before sea levels rose at the end of the last glacial period, and that they consistently used it over the last 13 000 years.[19] During this time sea levels were 85 m below present and the hill would have been about 80 km inland from the coast. Around 8000–9000 BP, sea levels around Australia rose, flooding the coastal plain and creating many islands including Tasmania and the Recherche Archipelago, and southern coastal landscapes as they appear today. Stories belonging to Australian Aboriginal groups tell of a time when the former coastline of mainland Australia was inundated by rising sea levels.[20]

Consequences

The original Sahul can be imagined as a land of bountiful ecosystems comprising mosaics of jungle, bushlands with forests, grasslands, rivers, wetlands and woodlands shaped by co-evolved plants, marsupials, birds and reptiles. They were populated by megafauna such as 200 kg grazing wombats, marsupial browsers such as the 1000 kg diprotodons and giant macropods, marsupial predators, *Genyornis* birds and emus, varanid lizards and a vast array of smaller animal species.[21] Their lack of experience with humans meant a wide variety of prey animals were naive to this new human predator, and the meat supply from these strange animals would have seemed unlimited to the first settlers. Conversely, many of the plants would have been familiar both generically and specifically, as they were present in much of South-East Asia. The adsorptive properties of clays and charcoal to neutralise plant toxins were well understood by people who already knew how to prepare a range of edible plant foods through heating, leaching and geophagy. As people moved out of tropical areas southwards new plant genera would have been encountered, requiring new knowledge development.[22]

The speed at which the inland parts of Australia were occupied implies rapid and expanding population growth from a large founding population. Genetic evidence supports as many as 1000 initial colonisers and, if they moved through the food-rich landscape as quickly as their ancestors had moved along the coast to Sahul, it may have taken as little as 1000–4000 years to settle the continent.[23] The abundance of food and favourable well-watered landscapes in the late Pleistocene are reflected in settlement patterns with plentiful evidence of early inland occupation, even in areas that are today classified as deserts. The inland regions received greater rainfall then than now and would have featured large permanent freshwater bodies, reliable riverine environments, tropical valleys and gorges. These vegetation patterns align with Anastasia Makarieva and Viktor Gorshkov's hypothesis, that precipitation does not depend entirely on distance from the coast, as illustrated by the Amazon and Yenisey river basins. In their hypothesis, a continuous forest from seashore to inland acts as a biotic pump, carrying plentiful precipitation inland independent of coastal distance. They further propose that destruction of a narrow band of forests around the continent's perimeter was sufficient to cause rainfall pattern change and drying of the interior of the Australian continent.[24] How this deforestation took place was perhaps a combination of human land use patterns, including cutting and burning the seemingly endless supply of trees to cook meat from the seemingly endless supply of animals, along with the cascading ecosystem changes, including wildfires, initiated by such human predatory behaviours.[4]

Aboriginal peoples living in the arid inland regions in historical times needed detailed in-depth food ecology knowledge and maintenance of complex near and far reciprocal social networks to survive in drought periods. A group with enough food would allow another group to enter their Country and share food resources in reciprocal arrangements. This sort of knowledge was unlikely to have been possessed by the original colonisers, so these adaptations developed later as the inland climate dried and food resources depleted as human populations grew and spread.[25]

Before human arrival, Australia was dominated by fire-sensitive plants that were managed by large and small fauna to create ecosystem heterogeneity and mitigate lightning-started wildfires through grazing patterns. Their rapid decline in abundance before actual extinction resulted in greater densities of ungrazed plant material and litter, which could then contribute to increased wildfires. This is likely to have caused a destructive fire cycle, possibly pushed along by human fire use that further contributed to extinction events and climatic changes, as the vital connections between coastal and inland forests declined. Research by Murray and Vickers-Rich supports this hypothesis, as the productive, relatively stable prehuman dry sclerophyll forests changed rapidly as fire increased soon after the first human colonisation around 60 000 BP.[26] This does not necessarily mean humans were consistently firing landscapes: after examining evidence from the fungus *Sporormiella* that depends on large herbivores to complete its life cycle, Susan Rule and co-researchers conclude that it is more likely that the changes were due to human hunting that caused rapidly diminishing populations of large herbivores. This led to relaxed grazing pressure and herbivore landscape management, and less control of lightning-started fire. This correlates with recent research in Africa which shows that as large herbivore populations (especially rhinos) dramatically decline through overhunting, grasses swiftly began to dominate, fire mitigation is lost and the proportion of forest containing fire-resistant sclerophyll plants increases at the expense of fire-sensitive rainforest.[27]

Gifford Miller and co-researchers have also established that around 40 000–45 000 BP there was a dramatic change in plant species distribution and abundance, including a shift from C4 grasses to C3 grasses.[iii] This would have led to further cascading changes to suites of smaller herbivores and their predators, and to changing fire regimes due to a build-up of flammable vegetation and replacement of fire-sensitive species with fire-adapted species.[28] It

---

iii The two photosynthetic pathways for plants to use carbon are C3 and C4. C3 is used by all trees, most shrubs, non-grass herbs and some temperate grasses. C4, evolved during the late Miocene when CO_2 levels were low, is used by most grasses, especially warm climate species including most native Australian perennial grasses, such as kangaroo grass (*Themeda triandra*), bandicoot and wallaby grasses (*Austrodanthonia* spp.)

would appear that the fire-resistant and dependent vegetation communities that dominate much of Australia today arose with the extinction of the herbivores that maintained the original ecosystems.

Trophic cascades: a domino effect

Kenneth McNamara and Peter Murray, who have participated in excavations at Madura Caves on the Nullarbor plain, have recovered mammal fossils dated from 16 000–38 000 BP. They include grey kangaroos, small predatory dasyurid marsupials, koalas and potoroos, and other higher rainfall species that prefer scattered forest ecosystems. This discovery indicates that the region formerly had a far moister climate than it does today. The megafauna appear to have largely disappeared, as they did elsewhere within 10 000 years of humans entering Australia.[29]

The decline of Australian megafauna has been further established using the presence or absence of fungal spores from fungal groups such as *Sporormiella*, which preferentially grow in the dung of large herbivores. High percentages of this fungus in lake and peat deposits are a good indicator of abundant megafauna herbivores. The fungal spores provide a measure of the presence or absence of megafauna in the fossil record and, as bio-indicators, show that these animals disappeared from fossil records around 50 000–40 000 BP.[30] This megafaunal decline reflects present situations in other parts of the world where keystone herbivore declines and extinctions lead to more homogeneous ecosystem states which are linked with secondary species extinctions, changes to community composition, and redefined carnivore guilds. Thus, the loss of only a few species – let alone over 90% – is likely to have resulted in far-reaching shifts in ecosystem resilience and may have promoted cascading shifts to irreversible new states.[31]

The consequences of species losses have been well researched overseas and provide models for what happened. In a domino effect, as one species drops out other connected and reliant species also disappear, and ecosystem collapse takes place as a spreading trophic cascade. An immediate outcome is discontinuation of efficient nutrient spread and cycling throughout landscapes from the more fertile hotspots into the less fertile parts.[32] A critical loss of large herbivore-dependent suites of dung beetle species, and their associated role in recycling herbivore dung, has been determined as affecting early Holocene European soils after megafauna extinction. No dung beetle species capable of dealing with large herbivore dung existed in Australia until introduced in 1974 and 1975 by CSIRO to deal with the 'new megafauna' dung from cattle, horses and other large introduced herbivores.[33,34] The long-term consequences of ending nutrient dispersion and incorporation by dung beetles are probably still being played out throughout most Australian soils (other than farmed grazing lands with introduced dung beetles), with ongoing declines in phosphorus levels and fertility. Research is needed on the roles that large introduced herbivores and a co-adapted dung beetle suite could have in Australian landscape health through nutrient dispersion and recycling.

Analysis from a cave site on the Nullarbor plains revealed that a giant wombat and 18 kangaroo species with a wide range of body sizes and morphologies, including two species of tree kangaroos (*Bohra* species), lived in the region during the middle to late Pleistocene. Also found were remains of bird species that required tree hollows for nesting.[35] This reflects a far more diverse landscape featuring a mosaic of grassy woodlands, shrublands and extensive C4 grasslands interspersed with many plants that had palatable fleshy leaves and fruit, and trees large enough for tree kangaroos. Today, the same area is primarily arid woodland dominated by mallee and morrel eucalypts and chenopod scrublands with low fire susceptibility, as the shrubs and trees are widely spaced with light grass cover that does not provide much fire fuel. The loss of mammal browsers, which possibly controlled mallee and morrel species and the saline soils they engineer, enabled these plants to expand their territories. Pollen histories show that C3 plants became the predominant group over C4 plants. The fact of their rise in abundance is supported by carbon isotope analysis of emu eggshells and wombat teeth, which indicates that plant communities, such as mallee woodlands, seem to have expanded in distribution and density after megafauna grazers disappeared.[30]

Plant and animal communities that co-evolved with large herbivores at different trophic levels would have disappeared. These include dung beetles, predators and scavengers that relied on large grazers for food, and plant species that relied on large grazers for seed dispersal. Quandong with bright red fruit attracted large birds (birds can see colour) as their seed distributors and survived because emus survived. South American researchers have identified that fruits attractive to mammals are characterised by a nutrient-rich pulp, grow low on the tree, are often dull coloured (as mammals see fewer colours than birds) and have indigestible nasty-tasting seeds with a tough endocarp. They have also shown that negative consequences for plant populations can be expected 'if the dispersal process is absent or impaired', with species either becoming extinct or having extremely restricted ranges in ecosystems without their co-evolved megafauna.[36] Some Australian plants retain obsolete defences against extinct browsers, such as thorns and spines which are common in the genera *Hakea, Acacia* and *Solanum,* all found in the Esperance bioregion and throughout Australia. Their defences may be useful again with the new 'megafauna' such as horses, camels, donkeys and cattle. Many smaller grazing and browsing species (and their predators) would have relied on the scleromorphic non-fire-resistant biodiverse woodlands with abundant food types and habitat mosaics created by the actions of larger grazing herbivores.[37] Such vegetation communities today are scattered remnants that have survived the extinction of megafauna, then Aboriginal people's burning practices and lastly the introduction of European land clearing practices and grazing animals.

There are several scenarios with various levels of evidence for causation of the extinction of over 90% of marsupial species following the first human invasion of Australia. One attributes all megafaunal extinctions to habitat and food loss caused by climate changes, perhaps enhanced by human landscape firing. This position dismisses the impact of humans as highly sophisticated invasive predators and implies a belief that Indigenous peoples lived in balance and harmony with nature. Other scenarios consider that Australia's complete lack of fossil megafauna kill sites and lack of archaeological evidence generally do not support the proposed human role in extinction. They also attribute extinctions to the inability of Australian fauna to adapt to climatic fluctuations.[38] Further arguments are made that humans were unlikely to have a role in megafauna disappearance during this time because a wide variety of smaller vertebrates, weighing less than 45 kg, also disappeared. Such assumptions reflect a lack of understanding of the tight interconnections within ecosystems, such that the loss of important keystone species can lead to extinction of co-adapted species. The outlook is further undermined by work from Mary Stiner and co-researchers in the Mediterranean. They have shown that Palaeolithic hunters increasingly targeted small rapidly reproducing game, after animals that were slower to reproduce, easier to catch and initially abundant, first declined then disappeared, along with their co-evolved cohorts, as human populations grew.[39] It seems that slower-reproducing small prey may have been a prerequisite to allow human populations to reach the critical densities that drive megafaunal extinction.[40]

These conclusions parallel modelling of various predation scenarios under a variety of environmental conditions by Steven Mithin for woolly mammoth extinction in the Russian plains and by Chris Johnson for diprotodon extinction in Australia. The models show that even relatively small increases in predation of less than 2% can cause population collapse and extinction.[41,42] Many slow-reproducing smaller species in Australia would have disappeared not only because of trophic collapses, but also due to human targeting and overexploitation as larger fauna became scarce. The faster-reproducing species such as kangaroos, wallabies, possums and bandicoots eventually became the dominant food supply. Human predation pressure was added to that of existing predator–prey relationships and, as a competitor, reduced the prey available to other predators.

There have been opinions that the first Australians would not have had the technology to hunt large herbivorous animals such as diprotodons, or predators such as marsupial lions (*Thylacoleo carnifex*) or giant monitor lizards (*Varanus priscus*).[40] The abilities of modern Indigenous hunters to kill large prey and predators invalidates this idea. An account by Jean Janmart of how Pygmy hunters in Africa could kill an elephant in minutes with only a knife and spears indicates how even an animal as large as a diprotodon could have been effectively dispatched.[43,iv] Taking eggs and hatchlings or targeting brooding parents of *Genyornis*, which co-existed with humans for only a short period, would have not been difficult. Like modern emus, they would have been most vulnerable when the males are sitting. Eggs can be stolen when the bird leaves the nest and personal experience with emus shows that the chicks can be caught relatively easily by running them into thick low scrub, where they become disoriented and tangled in the undergrowth.

Philip Clark describes how plant narcotics such as pituri (*Duboisia hopwoodii*) and corkwood (*Duboisia myoporoides*) were used to hunt kangaroos and emus in recent times.[44] Indigenous hunters would stupefy emus and fish by placing the crushed leaves of narcotic plants into waterholes. Strips of the bark were left by an affected waterhole to warn other humans. Methods such as this made hunting of large species, as well as birds, relatively easy. Animals that used dens in which to sleep and house their young (as many predators do) or that slept in hollows were also vulnerable to human hunters. Settler observations along the Murray River describe how Aboriginal possum hunters extracted their prey using a fur-twisting tool especially designed for this purpose.[45]

---

iv It took only a few minutes to bring an adult elephant down by hamstringing it.

The conclusions reached by those denying human culpability in Australian extinctions seem to underestimate human learning ability and development of hunting expertise. Richard Holdaway and Christopher Jacomb have researched Maori hunting of moa in New Zealand, and shown the expertise people can develop as they learn the most efficient ways to hunt different types of large and small prey.[46] Humans efficiently target the most obvious sources of animal food – large herbivores, and keystone predators considered dangerous and competitive such as *Thylacoleo* species. The targeting of juveniles of large animals can also effectively explain more drawn-out extinction events.[47]

Well documented research into similar but more recent impacts of human and non-human invasive predators in a naive prey situation illustrate that surplus killing (when a predator kills beyond its food needs, such as a fox does in a hen house) is also a characteristic of human behaviour.[48] Examples occurred in New Zealand, where human hunting and habitat destruction drove the 11 species of moa to extinction within 100 years of Polynesian settlement[46] and in North America, where humans shot so many passenger pigeons that the species went extinct.

The overall scenario would likely have contained elements of fast and slow extinction events occurring in parallel. Hiscock has suggested that the human populations fluctuated, and so might their relationships with ecosystems.[25] Ecologist Rosie Woodroffe has found that decline of large predators correlates closely with human population density and is worsened during periods of human population growth. The original colonisers were unlikely to have had intrinsic connections to the new land or an understanding of how the ecosystems worked and were interconnected.[49] When discussing the development of an ecological consciousness, Fred Kirschenman considers there is always a minority with foresight who can see serious future consequences if wasteful consumption practices are continued; however, like other past human societies and those of today, prevailing views trap societies into ignoring or being unwilling or unable to recognise shifting baselines.[50]

During the early years of settlement, people may have lived in large relatively permanent settlements as specialised big game hunters with hierarchical social structures, as is attributed to the development of agriculture. The Nawarla Gabarnmang rock shelter in the Kimberley, inhabited as long ago as 50 000 BP, may demonstrate this. The interior shows evidence of modifications and beautification with large well maintained frescoes upon the ceiling and walls.[51] Ian Keen and colleagues theorise that the supporting culture for such art was enabled by an easy availability of abundant food resources.[52] Not only human population growth but also settled human lifestyles have negative impacts on animals, particularly the killing of large predators for status reasons.[53]

Of the three original extinction scenarios presented above, the most likely seems to be a longer series of extinction events that reflected people moving into new areas, and human population fluctuations. The idea that humans had no impact is unsupported, but it appears that a fast and widespread blitzkrieg event was unlikely – though it happened to vulnerable species such as *Genyornis*. The middle scenario may be the best explanation. A combination of overhunting and elimination of keystone species as human populations grew and damaged pre-existing SES, led to a mix of fast and slow irreversible ecosystem changes, with human fire use triggering further ecological and climatic change and extinctions.

Afterwards: the landscapes of people

Where fire had been controlled by the dominant vegetation types and mosaics of large herbivore grazing patterns, with their demise infernos may have raged for days and weeks. This is occurring again in the huge unmanaged areas of parks, reserves and unallocated crown land in Western Australia and other regions of Australia.[54] At some period in the deep past, resilient food system solutions evolved as Aboriginal people accumulated knowledge. SES were developed and memories from Elders passed on that would allow them to not only fill the unoccupied habitat niches, but create new ones for themselves and surviving organisms. However, in doing so they reduced the potential future biodiversity, as other latent niche occupiers had nowhere for evolution to take them. Over a long period, ecosystems throughout Australia became realigned with Aboriginal systems of management that included sophisticated fire technology, cultivation techniques and cultural adaptations where the largest herbivores were kangaroo species. The loss of large animals involved the loss of their crucial role in maintaining landscape fertility via dung recycled into soils by dung beetles and soil organisms. This cannot be substituted by fire use.[34]

How people procured food influenced the adaptations of surviving animals and plants. For instance, kangaroos became smaller and attuned to people as predators as co-evolution took place.[55] Mallee eucalypts and proteaceous plants continued creating habitat for themselves and co-dependent species by engineering their soils. Small native mammals continued to be important in plant propagation, soil building, water balance and wildfire suppression as well as perpetuation of their food resources.[56] Some native plants, such as the Murnong daisy yam and youlk (*Platysace deflexa*) which produce an underground potato-like tuber, need major soil disturbance to regenerate. They likely evolved in concert with pig-like niche marsupials, such as the large extinct *Zygomaturus trilobus* and other digging species. With the survival of rabbit rats, corellas, bettongs and bandicoot species who had important soil-digging and disturbance behaviours when food foraging, these yam species persisted. They were further helped by humans with digging sticks disturbing the soil while looking for these plants as a food source. Other selection pressures were imposed on food and ecosystems through active management and distribution of food species into favourable environments by Aboriginal people.[57] Surviving burrowing animals, such as reptiles and ground-dwelling arthropods, also had ongoing influences on ecosystem soil structure, creating pores that allowed aeration and rainfall penetration and fertilising soils with their excreta.[58]

As discussed previously, plants can have profound effects on landscapes and ecosystems in their immediate habitats as well as wider climatic implications. Many of the key biophysical features of mallee and sandplain landscapes and ecosystems in Western Australia exist in their current form because of bioengineering processes – the dominant mallee/eucalypts create clay pavements and proteaceous plant species create laterite reefs. Both continued under Aboriginal SES.[59] Mallee is extremely resilient to disturbance, even by fire, as it regenerates rapidly from lignotubers to form the dominant canopy. Aboriginal fire regimes may have favoured its spread into new areas. In the past animal landscapes, large megafauna browsers such diprotodons and large macropods were possibly involved in patchily controlling this regrowth to allow other species to flourish, thus preventing the monocultural mallee communities of now.[60]

Humans are experts at altering and remaking ecosystems for their own benefit by capturing the largest proportion of total biotic energy. Aboriginal people became the ecosystem engineers and keystone species. Their activities shaped the new landscape regimes that were evident when Europeans first arrived in Australia.[61,62] Aboriginal Dreamtime stories reflect these changes as well as geophysical events, with descriptions of environmental happenings from the deep past. A local Esperance bioregion Nyungar and Ngadju traditional story, 'The Story of Jimbalana', may reflect traditional ecological knowledge for it seems to be a warning about human damage and environmental change enacted by spirits. The story describes how the Esperance bioregion salt lakes, which the story describes as originally fresh, came into being after Jimbalana cleared large areas of bush with his boomerang. The story demonstrates ecological knowledge linking land clearing to salinity.[63] Tales such as these are likely some of the longest continuous record of historic eco-geological events in the world. They play important roles in presenting memories of past events as well as acting as teaching models that ensure accurate transmission of ecological, cultural and technical knowledge and skills to future generations.[64]

Eventually the early, poorly adapted SES of the ancestral peoples would have been replaced by ways of living that were more in tune with the short- and long-term environmental variability that they had inadvertently contributed to in the Australian landscape. The 50+ millennia of human occupation in Australia is such a long period that it allows for many fluctuations of population growth and intensification of land use.[27]

The mid to late Holocene was a period of population growth for Aboriginal people despite El Niño events causing increased climate variability, for there is evidence of greater environmental impacts as resource use intensified.[65] Bradley Smith and Carla Litchfield conclude that during this period there were multiple interactions with non-Indigenous peoples from 3100–3400 BP. Dingoes that had spent time in Australia and then returned to Asia had picked up the marsupial biting lice (*Heterodoxus spiniger*) which subsequently spread to canids worldwide, excepting in northern latitudes. At this time the dingo (*Canis dingo*) became established in Australia, perhaps coming in with Dravidian settlers of Indian origin around 4000 years BP. Dingoes are implicated in the demise of the thylacine (*Thylacinus cynocephalus*) and Tasmanian devil (*Sarcophilus harrisii*) on mainland Australia, before they became a functioning and integral member of Australian ecosystems. The Dravidian settlers may have introduced cultural and technological changes that disrupted previous cultural mores that controlled population growth and ecological impacts, including intensified plant cultivation, and use of dingoes to increase hunting efficiency. James Boyce describes how the introduction of dogs into Tasmania completely transformed the ability of Aboriginal societies to catch kangaroos and other prey, to a greater extent than guns.[66,67] Dingoes also became a new food source, which effectively controlled their population levels and reduced their impacts on indigenous fauna and predatory competition with Aboriginal hunters.[68,v]

Some researchers attribute the period of intensification during the Holocene as evidence supporting theories of cultural progression towards greater complexity and intensification into agriculture. Yet, if looked at from the very long view, it may simply reflect another

---

v   Tom Dimer (note 67) describes hunting with dingoes.

cultural adaptive cycle in a continent occupied for over 50 000 years, rather than linear progressions towards agriculture.[25]

There have been three principal extinction periods in south-western Australia. The first, described above, took place around 40 000–50 000 BP (no sites in Western Australia are currently dated beyond this) when people first invaded and settled. The second period in the lower south-west started during the 15 000 years before European arrival and intensified after the introduction of the dingo. Several wallaby species, hopping mice, a bat species, long nosed bandicoots (*Perameles* spp.), Tasmanian devils (*Sarcophilus*) and eventually the thylacine (*Thylacinus*) disappeared from the mainland.[29] The third extinction period in Australia started with European invasion and settlement, and continues today. For instance, prior to European settlement in the Esperance bioregion in 1863, there were significant populations of wallaby species including the burrowing bettong (*Bettongia lesueur*), tammar (*Macropus eugenii*), woylie (*B. penicillata*), the banded hare wallaby (*Lagostrophus fasciatus*), rufous hare wallaby (*L. hirsutus*), crescent nail-tail wallaby (*Onychogalea lunata*) and black gloved wallaby (*M. irma*).[69] Most of these are now gone.

Ecology in culture

Anthropologist Gregory Cajete describes how ecology was an intrinsic part of Native American traditional education. This seems to be a feature of most indigenous cultures who demonstrate a contiguous set of cultural structures (concepts, metaphors and language) that describe in detail their linkages to nature.[70,71] In discussing the Aboriginal Dreaming concept, Aboriginal author Nancy Williams points out that despite the concept interpreted as universal by non-Aboriginal people, it is highly influenced by local environments. However, there is consensus that all Aboriginal groups believe their existence is embedded in and connected to the rhythms of Country where ancestral mythical beings and spirits travelled along specific pathways as they created the biophysical reality.[72] All of Aboriginal Australia was linked through songlines, the geographical paths along which the creator ancestors moved to bring the present landscape into being. There is a songline that connects the Kimberley to the Centre to Port Augusta and west through the Esperance bioregion to Albany. This belief system, which so intimately entwines people and nature, allows for local environmental variations but ensures that management systems followed similar models everywhere.[73] This is supported by Edward John Eyre's observation in 1840:

> As far as has yet been ascertained, the whole of the aboriginal inhabitants of this continent, scattered as they are over an immense extent of country, bear so striking a resemblance in physical appearance and structure to each other; and their general habits, customs, and pursuits, are also so very similar, though modified in some respects by local circumstances or climate …[74]

The early explorers' portrayals of the southern parts of the Esperance bioregion describe people living comfortably in a managed and productive landscape. In the sandplain and parts of the mallee, the people had developed SES as the keystone culture based on a pastoral economy of ecoagriculture. Along the coastal region there were abundant freshwater or only slightly brackish swamps, and water in rockholes, many springs and river pools. Granite domes dotted the landscape, providing fertile soil oases around their margins from decomposing granite and the water runoff creating damp microclimate habitats. There were numerous mosaics on the better yate soils of grassland feed maintained by Aboriginal people who burnt patches for grassland species. The habitats of smaller mammals, including wallaby and possum species, birds, reptiles and invertebrates, were directly benefited by the heterogeneous vegetation communities that encompassed a proliferation of edge effects of grassland and bush, woodlands and denser forest and thickets. The constant and purposeful anthropogenic disturbance resulted in a diversity of niches creating abundant biodiversity, increasing the range of food and habitat resources available for animals, plants and people through space and time.[62,75]

Perhaps the Dreaming stories passed on to each generation were a cultural method of redressing the earlier maladaptation and outlining adaptive pathways to follow. This capacity may have stabilised SES for later Aboriginal people as they learnt to manage what their forebears had left, in a relatively resilient and sustainable manner.

Endnotes

1. Konnor M (1990) *Why the Reckless Survive: and Other Secrets of Human Nature*. Viking, New York.
2. White L (1967) The historical roots of our ecologic crisis. *Science* **155**(3767), 1203–1207. doi:10.1126/science.155.3767.1203
3. Flannery T (2001) *The Eternal Frontier: An Ecological History of North America and Its Peoples*. Text Publishing, Melbourne.
4. Flannery T (1994) *The Future Eaters: An Ecological History of Australasian Lands and People*. Reed Books, Sydney.
5. Oppenheimer S (2014) Modern human spread from Aden to the Antipodes with passengers and when? In *Southern Asia, Australia and the Search for Human Origins*. (Eds R Dennell, M Poor) pp. 228–242. Cambridge University Press, Cambridge.
6. Johnson CN (2009) Ecological consequences of Late Quaternary extinctions of megafauna. *Proceedings. Biological Sciences* **276**, 2509–2519. doi:10.1098/rspb.2008.1921
7. Silliman BR, McCoy MW, Angelini C, Holt RD, Griffin JN, van de Koppel J (2013) Consumer fronts, global change, and runaway collapse in ecosystems. *Annual Review of Ecology Evolution and Systematics* **44**, 503–538. doi:10.1146/annurev-ecolsys-110512-135753
8. Kull K (2014) Adaptive evolution without natural selection. *Biological Journal of the Linnaean Society* **112**(2), 287–294. doi:10.1111/bij.12124
9. Rose DB (1996) *Nourishing Terrains: Australian Aboriginal Views of Landscape and Wilderness*. Australian Heritage Commission, Canberra.
10. Monash University (2014) *Explore Sahul Time*. <http://sahultime.monash.edu.au/explore.html>
11. Cooper A, Stringer CB (2013) Did the Denisovans cross Wallace's Line? *Science* **342**(6156), 321–323. doi:10.1126/science.1244869
12. Clarkson C, Jacobs Z, Marwick B, Fullagar R, Wallis L, Smith M, Roberts RG *et al.* (2017) Human occupation of northern Australia by 65,000 years ago. *Nature* **547**(7663), 306–310. doi:10.1038/nature22968
13. Turney CSM, Bird MI, Fifield LK, Roberts RG, Smith M, Dortsch CE, Grun R *et al.* (2001) Early human occupation at Devil's Lair, southwestern Australia 50,000 years ago. *Quaternary Research* **55**(1), 3–13. doi:10.1006/qres.2000.2195
14. Gillespie R (2004) First and last: dating people and extinct animals in Australia. *Australian Aboriginal Studies* **1**, 97–104.
15. Pugach I, Delfin F, Gunnarsdóttir E, Kayser M, Stoneking M (2013) Genome-wide data substantiate Holocene gene flow from India to Australia. *Proceedings of the National Academy of Science* **110**(5), 1803–1808.

16. Malaspinas AS, Westaway MC, Muller C, Sousa VC, Lao O, Alves I *et al.* (2016) A genomic history of Aboriginal Australia. *Nature* **538**(7624), 207–214. doi:10.1038/nature18299
17. Stoneking M, Kraus J (2011) Learning about human population history from ancient and modern genomes. *Nature Reviews. Genetics* **12**(September), 604–614.
18. Flood J (2006) *The Original Australians: Story of the Aboriginal People*. Allen and Unwin, Sydney.
19. Smith M (1993) Recherché a l'Esperance. PhD thesis, Department of Anthropology, University of Western Australia, Perth.
20. Nunn P, Reid JN (2016) Aboriginal memories of inundation of the Australian coast dating from more than 7000 years ago. *Australian Geographer* **47**(1), 11–47. doi:10.1080/00049182.2015.1077539
21. Owen-Smith N (2013) Contrasts in the large herbivore faunas of the southern continents in the late Pleistocene and the ecological implications for human origins. *Journal of Biogeography* **40**, 1215–1224. doi:10.1111/jbi.12100
22. Rowland MJ (2002) Geophagy: an assessment of the implications for development of Australian Indigenous plant processing technologies. *Australian Aboriginal Studies* **1**, 58.
23. Williams AN (2013) A new population curve for prehistoric Australia. *Proceedings of the Royal Society B, Biological Sciences* **280**(1761), 5–8.
24. Makarieva AM, Gorshkov VG (2006) Biotic pump of atmospheric moisture as driver of the hydrological cycle on land. *Hydrology and Earth System Sciences Discussions* **3**(4), 2621–2673. doi:10.5194/hessd-3-2621-2006
25. Hiscock P (2008) *Archaeology of Ancient Australia*. Routledge, London.
26. Murray PF, Vickers-Rich P (2004) *Magnificent Mihirungs: The Colossal Flightless Birds of the Australian Dreamtime*. Indiana University Press, Bloomington.
27. Rule S, Brook BW, Haberle SG, Turney CSM, Kershaw AP, Johnson CN (2012) The aftermath of megafaunal extinction: ecosystem transformation in Pleistocene Australia. *Science* **335**(6075), 1483–1486. doi:10.1126/science.1214261
28. Miller GH, Fogel ML, Magee JW, Gagan MK, Clarke SJ, Johnson BJ (2005) Ecosystem collapse in Pleistocene Australia and the human role in megafaunal extinction. *Science* **309**(5732), 287–290. doi:10.1126/science.1111288
29. McNamara K, Murray P (2010) *Prehistoric Mammals*. Western Australian Museum, Perth.
30. van der Kaars S, Miller GH, Turney CSM, Cook EJ, Nürnberg D, Schönfeld J *et al.* (2017) Humans rather than climate the primary cause of Pleistocene megafaunal extinction in Australia. *Nature Communications* **8**, 14142. doi:10.1038/ncomms14142
31. Holling CS, Gunderson LH, Peterson GD (2002) Sustainability and panarchies. In *Panarchy: Understanding Transformations in Human and Natural Systems*. (Eds LH Gunderson, CS Holling) pp. 63–102. Island Press, Washington, DC.
32. Doughty CE, Wolf A, Malhi Y (2013) The legacy of the Pleistocene megafauna extinctions on nutrient availability in Amazonia. *Nature Geoscience* **6**, 761–764. doi:10.1038/ngeo1895
33. Svenning JC (2002) A review of natural vegetation openness in north-western Europe. *Biological Conservation* **104**, 133–148. doi:10.1016/S0006-3207(01)00162-8
34. Nichols E, Spector S, Louzada J, Larsen T, Amezquita S, Favila ME (2008) The Scarabaeinae research network: ecological functions and ecosystem services provided by Scarabaeinae dung beetles. *Biological Conservation* **141**, 1461–1474. doi:10.1016/j.biocon.2008.04.011
35. Prideaux GJ, Long JA, Ayliffe LK, Hellstrom JC, Pillans B, Boles WE *et al.* (2007) An arid-adapted middle Pleistocene vertebrate fauna from south-central Australia. *Nature* **445**(7126), 422–425. doi:10.1038/nature05471
36. Guimarães PR, Galetti M, Jordano P (2008) Seed dispersal anachronisms: rethinking the fruits extinct megafauna ate. *PLoS One* **3**(3), e1745.
37. Owen-Smith N (2013) Contrasts in the large herbivore faunas of the southern continents in the late Pleistocene and the ecological implications for human origins. *Journal of Biogeography* **40**, 1215–1224. doi:10.1111/jbi.12100
38. Wroe S, Field J, Fullagar R, Jermin LS (2004) Megafaunal extinction in the Late Quaternary and the global overkill hypothesis. *Alcheringa: An Australasian Journal of Palaeontology* **28**(1), 291–331. doi:10.1080/03115510408619286

39. Stiner MC, Munro ND, Surovell TA, Tchernov E, Bar-Yosef O (2004) Palaeolithic growth pulses evidenced by small mammal exploitation. *Science* **283**(January), 190–196.
40. Bulte E, Horan RD, Shogren JF (2006) Megafauna extinction: a paleoeconomic theory of human overkill in the Pleistocene. *Journal of Economic Behavior & Organization* **59**(3), 297–323. doi:10.1016/j.jebo.2005.04.010
41. Mithin S (2003) *After the Ice: A Global Human History 20,000–5,000 BC*. Orion Books, London.
42. Johnson CN (2006) *Australia's Mammal Extinctions: A 50,000-year History*. Cambridge University Press, Melbourne.
43. Janmart J (1952) Elephant hunting as practiced by Tee Congo pygmies. *American Anthropologist* **54**(1), 146–147. doi:10.1525/aa.1952.54.1.02a00440
44. Clarke PA (2011) *Aboriginal People and Their Plants*. Rosenberg Publishing, Sydney.
45. Broome R (2018) Working ancient mallee lands. In *Changing Landscapes, Changing People: Australia's Southern Mallee Lands, 1830 –2012*. (Eds A Gaynor, K Holmes, R Broome, C Fahey, R Ford). La Trobe University, Melbourne.
46. Holdaway RN, Jacomb C (2000) Rapid extinction of the moas (Aves: Dinornithiformes): model, test and implications. *Science* **287**(5461), 2250–2254. doi:10.1126/science.287.5461.2250
47. Brook BW, Johnson CN (2006) Selective hunting of juveniles as cause of the imperceptible overkill of the Australian Pleistocene megafauna. *Alcheringa: An Australasian Journal of Palaeontology* **30**(S1), 39. doi:10.1080/03115510609506854
48. Short J, Kinnear JE, Robley A (2002) Surplus killing by introduced predators in Australia: evidence for ineffective anti-predator adaptations in native prey species? *Biological Conservation* **103**, 283–301. doi:10.1016/S0006-3207(01)00139-2
49. Woodroffe R (2000) Predators and people: using human densities to interpret the decline of large carnivores. *Animal Conservation* **3**, 165–173. doi:10.1111/j.1469-1795.2000.tb00241.x
50. Kirschenman F (2010) *Cultivating an Ecological Conscience: Essays from a Farmer Philosopher*. Kentucky University Press, Lexington.
51. Delannoy JJ, Bruno D, Geneste JM, Katherine M, Barker B, Whear RL, Gunn RG *et al.* (2013) The social construction of caves and rockshelters: Chauvet Cave (France) and Nawarla Gabarnmang (Australia). *Antiquity* **87**(335), 12–29. doi:10.1017/S0003598X00048596
52. Keen I, Arnold J, Feinman G, Hayden B, Kelly R, Peterson N *et al.* (2006) Constraints on the development of enduring inequalities in Late Holocene Australia. *Current Anthropology* **47**(1), 7–38. doi:10.1086/497672
53. Ripple WJ, Estes JA, Beschta RL, Wilmers CC, Ritchie EG, Hebblewhite M *et al.* (2000) Status and ecological effects of the world's large carnivores. *Science* **343**(6167), 1241484.
54. Tiedeman K (2011) District Manager, Department of Conservation, Esperance. Pers. comm.
55. Codding BF, Bird RB, Kauhanen PG, Bird DW (2014) Conservation or co-evolution? Intermediate levels of Aboriginal burning and hunting have positive effects on kangaroo populations in Western Australia. *Human Ecology* **42**(5), 659–669. doi:10.1007/s10745-014-9682-4
56. Martin G (2003) The role of small ground-foraging mammals in topsoil health and biodiversity: implications to management and restoration. *Ecological Management & Restoration* **4**(2), 114–119. doi:10.1046/j.1442-8903.2003.00145.x
57. Goodall G (2014) Botanist. Interviewed by Nicole Chalmer, Albany, 17 July.
58. Abbott I, Parker CA (1980) Agriculture and the abundance of soil animals in the Western Australian wheatbelt. *Soil Biology & Biochemistry* **12**, 455–459. doi:10.1016/0038-0717(80)90028-0
59. Verboom W, Pate JS (2013) Exploring the biological dimensions to pedogenesis with emphasis on the ecosystems, soils and landscapes of southwestern Australia. *Geoderma* **211–212**, 154–183. doi:10.1016/j.geoderma.2012.03.030
60. Pate J (2014) Emeritus Professor of Botany. Interviewed by Nicole Chalmer, Denmark, 6 November.
61. Smith EA, Wishnie M (2000) Conservation and subsistence in small-scale societies. *Annual Review of Anthropology* **29**(1), 493–524. doi:10.1146/annurev.anthro.29.1.493
62. Bird RB, Tayor N, Codding BF, Bird DW (2013) Niche construction and Dreaming logic: Aboriginal patch mosaic burning and varanid lizards (*Varanus gouldii*) in Australia. *Proceedings of the Royal Society B. Biological Sciences* **208**(1772), 20132297.
63. Graham S, Castletown Primary School Children (2008) *The Story of Jimbalana*. PALS Reconciliation Project.

64. Sveiby K, Skuthorpe T (2006) *Treading Lightly: The Hidden Wisdom of the World's Oldest People*. Allen and Unwin, Sydney.
65. Johnson CN, Brook BW (2011) Reconstructing the dynamics of ancient human populations from radiocarbon dates: ten thousand years of population growth in Australia. *Proceedings of the Royal Society B. Biological Sciences* **278**(1725), 3748–3754.
66. Hassell E (1936) *My Dusky Friends*. C.W. Hassell, Perth. Reprinted 1975.
67. Hassell E, Davison DS (1936) Notes on the ethnology of the Wheelman tribe of south-western Australia. *Anthropos*, 679–711.
68. Dimer T (1990) *Outback Station Life; Aboriginal Customs and Beliefs; Bush Skills and Survival; Vermin Control for Agricultural Protection Board*. OH 2339. Interviewed by Helen Crompton (transcription). Oral History Unit, J.S. Battye Library, Perth.
69. Abbott I (2009) *Faunal Extinctions: Where and How have Populations Disappeared?* Information Sheet 21. Science Division, Department of Environment and Conservation, Perth.
70. Cajete G (1994) *Look to the Mountain: An Ecology of Indigenous Education*. Kivaki Press, Durango.
71. Prober SM, O'Connor MH, Walsh FJ (2011) Australian Aboriginal peoples' seasonal knowledge: a potential basis for shared understanding in environmental management. *Ecology and Society* **16**(2), 12. doi:10.5751/ES-04023-160212
72. Williams NM (1986) *The Yolngu and Their Land: A System of Land Tenure and the Fight for its Recognition*. Stanford University Press, Palo Alto.
73. Gammage B (2011) *The Biggest Estate on Earth: How Aborigines made Australia*. Allen and Unwin, Sydney.
74. Eyre EJ (1845) *An Account of the Manners and Customs of the Aborigines and the State of Their Relations with Europeans*. T. and W. Boone, London.
75. Ries L, Fletcher JR, Battin RJ, Sisk TD (2004) Ecological responses to habitat edges: mechanisms, models, and variability explained. *Annual Review of Ecology Evolution and Systematics* **35**, 491–522. doi:10.1146/annurev.ecolsys.35.112202.130148

4
How to sustain eating nature

> Not a wilderness, not a land peopled by wanderers, but a managed landscape created by the enormous labour [and intellect] of a people intent on creating the best possible conditions for food production.[1]

There is a still persistent belief that Aboriginal people's relationship with the land was as simple hunter–gatherers, wandering through some primeval Australian Eden as passive recipients of whatever foods their environment had to offer and playing little role in deliberately managing land, except for a few fires. Early observations by European/Anglo explorers and settlers do not support this assumption. In 1790, First Fleet Captain of the *Sirius* and artist John Hunter showed an understanding that Aboriginal people used fire for clearing land, and in 1848, Major Thomas Mitchell famously wrote that 'fire, grass, kangaroos, and human inhabitants … [were] … all dependent on each other for existence'.[2,3] In this chapter I explore the relationships that Aboriginal peoples had with the land to produce food, as they used a long series of historical learnings and ecological knowledge to culturally manage food-producing systems.

Humans can very easily overexploit the edible resources in a landscape, so for long-term sustainability self-regulatory mechanisms of restraint need to be embedded in a culture to protect both individual plant and animal food species and the ecosystems they are part of. In his classic research on religion and food production in traditional societies, Eugene Anderson discusses principles of resource conservation through using cultural mores, spirituality and religion to deliberately protect areas in Indigenous societies.[4] This is supported by Tim Flannery's observations during field work in New Guinea during the 1980s and 1990s, where he describes how religious taboos were used to permanently protect breeding reserves. It was only overspill that could be taken for food.[5] This characteristic has been found in all successfully sustainable long-term Indigenous cultures. Along with food taboo systems and restrictions to population growth, it allowed species and ecosystems to escape continuous human predation and overconsumption.[6]

We as humans seem to have maladaptive genetic imprints that are the seeds of environmental destruction and therefore our own demise, unless constrained by powerful enduring cultural mores. In his examination of how present environmental problems are likely to have evolutionary linkages, evolutionary biologist Dustin J. Penn proposes that a major explanatory factor in human overexploitation behaviour is found in the concept of 'socially conspicuous consumption'. This involves culturally endorsed overkill and resource overexploitation. The dynamics of this concept propose that, within hierarchical systems, short-term social and sexual status gains for individuals or groups can be achieved by demonstrating the ability to be wasteful, using resources in a display of profligate wealth. Penn links development of hierarchies and their associated inequalities to increasing population density, sedentarism and reduced culture–nature linkages as food storage improves and food production systems intensify.[7]

Karl-Erik Sveiby, a management consultant, and Tex Skuthorpe, a Nhunggabarra man from New South Wales, suggest that, in contrast, Aboriginal social organisations evolved to deliberately avoid hierarchical systems. Instead, they used collective leadership systems, both within and between sexes, and prevented reckless overuse of resources by controlling population growth and status-seeking behaviour.[8] Various anthropological and human ecology researchers also describe how most Indigenous societies have ecological models and environmental behaviours integrated with moral and religious beliefs, allowing knowledge, practice and beliefs to co-evolve and be culturally passed down to successive generations, avoiding hierarchical practices such as class systems, and a judicial system.

Dame Mary Gilmore, born in 1865 at Wagga Wagga in New South Wales, grew up in close association with local Aboriginal groups there and around south-western New South Wales, including Houlaghans Creek and the Murrumbidgee and Lachlan river systems. In her 1934 book *Old Days: Old Ways*, she explains that Aboriginal societies had teaching systems and were highly educated – though not necessarily in the same way as Europeans interpreted education. Knowledge level status in Aboriginal society had to be earnt through a system of stepwise education, and was not inherited. Not everyone was capable of achieving the same level of knowledge.[9] This embodied cultural learning enabled the ability to adapt to changing environmental conditions rather than rely on past rigid hierarchies to pass on knowledge.

Anthropologists Eric Smith and Mark Wishnie, who studied the resource usage of hunter-gatherers in Amazonia, also found that resources were managed to prevent tribal overexploitation. A variety of systems were used, including socially regulated access, management rules governing resource harvesting, breeding reserve systems, means to monitor compliance with the cultural rules and methods to punish those who violated them.[10] Observations by explorers and settlers in the Australian context supported these strategies. Consciously or intrinsically, Aboriginal people used methods embedded in culture to control their impacts on the long-term food-producing potentials of their landscapes. Sometimes people consciously understood the reasons for particular strategies, others have become so embedded in religion and culture the original reasons were lost.

The ecological roles that Aboriginal people displayed were extremely important to the social ecological systems (SES) of many animal species they relied upon as food. As the keystone species/culture and ecosystem engineers, Aboriginal peoples in Australia created and maintained ecological habitats and niches primarily with judicious fire use. The discontinuation of their role has been viewed as the most important cause of the cascading extinctions of small mammals in arid areas. New research by anthropologist Brian Codding and co-researchers supports this view, for it shows that where species have a long history of consistent interactions with people, they have become co-evolved to human disturbance and predation and need this continued interaction to successfully persist.[11]

Aboriginal SES, at the time of European contact, generally displayed a suite of culturally embedded adaptive behaviours that ensured a culture of restraint along with systems of food storage and various levels of settlement and mobility. The restraint behaviours seemed to have evolved as definitive cultural mechanisms to perpetuate food security. These aspects of the SES may have developed initially as a reaction in the deep past to warfare events and starvation triggered by unsustainable past resource use. As well, there were other periods of food shortage including when sea levels rose in the early Holocene around 10 000 years ago, forcing coastal peoples into smaller territories. These cultural adaptations (the level of genetic or epigenetic adaptation involved is unexplored) effectively empowered Aboriginal SES to withstand the times of shortage that periodically arose, including those caused by long-term climatic fluctuations described for the Western Australian south coast by Louise Cullen and Pauline Grierson.[12]

Hunter–gathering or ecoagriculture?

There are so many variabilities in how people who have been categorised as hunter-gatherers interact with their environment ecologically and culturally, that no one definition can fit all. This term tends to homogenise the lifestyles of people who lived in so-called recognisably 'non-agricultural' societies. This way of life has historically and even in present times been judged as being 'lower' on some evolutionary cultural scale – either as living in the primeval state of humanity that 17th-century philosopher Thomas Hobbes depicted as 'solitary, poor, nasty, brutish and short'; or as highly influential English poet John Dryden (also 17th century) described, living as 'noble savages' – a state from which the rest of humanity had fallen. The first was used by Europeans to justify improving Indigenous people's ways of life by invasion and colonisation; the other as a requiem to a doomed way of life to be eased in its passing.[13]

Despite the variability in ways of living, researchers who believe in the subsistence model endorse it as an important common factor in hunter-gatherer lifestyles that is claimed to have characterised human existence for 99% of prehistory.[13] They also consider that although hunter-gatherers both influenced their environments and were in turn impacted by them, their lifestyles were characterised by food and other resources provided by unmanaged wild plant and animal communities, with a lack of animal and plant domesticates. Researchers in the Department of Anthropology and Ecology at the University of California have cited some Australian Aboriginal groups as examples of contemporary hunter-gatherers who lived with an ecological economy.[14] For example, when modelling random foraging behaviour in productive mosaics, 'a forager nearly always will depart a patch before it has been fully depleted of resources'. The breeding stock left allows the patch to recover. This is described as an unplanned occurrence, implying that hunter-gatherers did not consciously plan for recovery. Such an approach is highly questionable in terms of its applicability to Australian Aboriginal people (or any Indigenous peoples), as it precludes learning, choices and the passing on of intergenerational ecological knowledge about human environmental impacts on food production. There are numerous examples in Australia where this is not the model for hunter-gathering, as the depth and extent of landscape management moved food production beyond subsistence into systems that ensured high degrees of certainty and abundance. Evidence of deliberate plantings and gardening of various plant food species, as well as purposeful fire use, also indicates a significant role in land management and niche construction.

Within the range of human food-producing activities, I propose that Aboriginal people at the time of colonisation practised ecoagriculture. They managed fire in sophisticated ways to create and maintain habitats and ecosystems. They stored some food with simple technologies, such as burying in pits, storing grass seed in tightly woven baskets, grinding and mixing with tree gum into balls that could be rejuvenated by adding water, and cooked fish that had been stored by wrapping in paperbark. Their primary technique, though, used ecological principles acquired by qualitative observation of habitat preferences, life histories and behaviour patterns of animal and plant food species, to create and maintain habitats as a managed living food surplus.[15] They were farmers and food gardeners, for though animals and plants may not have been domesticated in the generally accepted sense, many were as co-evolved as are traditional domesticates, and extremely reliant upon Aboriginal management techniques for their well-being and survival. Since many of the animals and plants were also ecosystem engineers, these systems reflected a mutualistic interdependent web of relationships that established resilient human and non-human SES to maintain ecosystems. Aboriginal ecoagriculture was an extreme example of what American farmer-philosopher Fred Kirschenman says is vital for sustainable agriculture – keeping 'the wild' biodiversity in human managed systems, which therefore benefits both people and nature as natural processes continue.[16]

Managing people

Population control systems as an institutionalised cultural strategy had the largest impact on allowing Aboriginal people to sustain their food production systems for future generations. Though long-term population levels may have fluctuated with food availability, there is evidence that Aboriginal SES deliberately managed their populations.

In Aboriginal groups, there were a variety of cultural beliefs that limited reproduction rates to replacement rather than growth. Some of these methods may seem repugnant to a modern urban society that is disconnected from nature, but they reflect the harsh reality that what is beneficial for a group's survival through millennia of fluctuating environmental resources, is not necessarily pleasant at the individual level. They included direct and indirect actions that reduced the fertility and therefore number of people in any given group. Though long-term population levels may have fluctuated with food availability, there is abundant evidence from many early explorers' descriptions throughout Australia of a landscape with abundant wildlife and fish. In his 1841 journal reports when surveying the southern coast of Western Australia, surveyor John Septimus Roe noted the abundance of kangaroos and emus along the coastal mosaic of grasslands and bushlands of the Esperance bioregion.[17] These animals were associated with extensive semi-permanent Aboriginal settlements, and their common occurrence indicated a population living below carrying capacity.

Modern western societies use various methods for birth control, including birth control pills or implants, the 'morning after' pill, abortion, condoms, vasectomy and (with variable success) abstinence from sex at critical ovulation periods. These choices are important to women's rights to manage their individual sexual and reproductive matters, as well as family planning, though rarely are they linked to any long-term population growth planning policies.[18] Aboriginal peoples also had methods of reproductive control and family planning, but these were employed within the context of how many people their environment could safely support in bad years. The number of children born was extremely important to them, not only as individuals but also as a society intimately connected to their ecosystem. Importantly, it should be noted that their traditional societies seemed far better and more advanced than modern societies in understanding that population size does matter in long-term environmental sustainability.[19,i]

This limiting of population was the most important practice that allowed Aboriginal people to have indefinite food and resource sustainability. In terms of childbearing, Aboriginal women at the time of European invasion were better off than most European women as they were not expected to have numerous children year after year until physically worn out. Ethel Hassel, an early 1870s settler in Jarramongup, observed this for the local Wheelman group, 'They did not have very large families, and there were long intervals between each child [of 4–5 years].'[14] This pattern is recorded as being practised in other places in Australia and the world. Environmental activist and farmer Wendall Berry reports that the Hunza of northern Pakistan practised sexual restraint with births four to five years apart as a form of birth control, and that the practice only declined with the coming of western civilisation and its false promises of endless food and resources.[20]

Other methods included cultural eating practices that result in extreme leanness in women, thus affecting fertility (irregular menses were apparently the norm), augmented by male subincision, contraception and abortions (certain plants could be eaten) as well as trauma.[21] Subincision was a method to reduce fertility in men. It carried significant cultural kudos among men, with only married men appearing to have undergone it.[22,23] Intergroup warlike behaviour inadvertently controlled population. There were revenge and payback killings, that eminent anthropologist Henry Reynolds describes as the major cause of death by violence in Aboriginal societies. Endemic customary institutions and practices related to kinship bonds and classificatory relationships fostered feuds but also inhibited their escalation into the type of mass warfare that has killed so many millions in western and other societies.[24] Feuding was so common in the Albany Nyungar tribes that Isaac Nind, the garrison doctor for Albany during 1826–1829, commented that 'They are so constantly at war that their numbers must be considerably diminished by it.[ii] When an individual falls, there is always some who take it upon themselves to revenge his death'.[25] In his respected book *Broken Spears*, Neville Green notes that in the first 25 years of European settlement feuding was observed and recorded by several writers as an ongoing practice. Even before European contact, 'this custom must have had a marked impact on the tribal population'.[26]

Managing nature

As well as population management, environmental practices included rotational resource use derived from mobility systems that reduced harvest pressure at pre-determined times with

---

i   Bignell describes local Aboriginal people saying to settlers how important it was to control the number of people living in an area.
ii His journals are recognised ethnographically as a vital important source of observations on Nyungar lifeways in Western Australia, pre-European settlement.

periods of rest, so allowing populations of food species time to recover. Archaeologist Miles Mitchell provides evidence that Esperance bioregion Nyungar people combined a largely sedentary occupation of the coastal lands with periodic seasonal trips further inland, linked to rotational resting strategies for food animals and plants along with seasonal availability.[27] These in turn are dependent upon human population size and animal/plant food biomass, that are fundamentally linked to landscape mosaics of soil nutrients and water. The highly mobile lifestyle of people of the arid Nullarbor plains (compared to the Esperance bioregion) reflected a lower food biomass that was more ephemeral and seasonally related, so people had to move regularly to prevent serious overexploitation.[28]

The culturally embedded and transmitted mechanism of seasonal calendars described by the Ngadju people to CSIRO's Susan Prober and colleagues, provided the core knowledge-based restraint frameworks for food resource management. Their calendar had two major seasons and four sub-seasons while the more coastal Nyungar calendar had six main seasons. Ngadju Country encompasses the northern woodlands and mallee, as well as coastal sandplain around Point Malcolm east to Point Culver in Western Australia. There were and still are rules and prescriptions determining the seasonal timing, frequency, intensity and long-term patterns of usage. This enabled a resilient and diverse resource base to be sustained.[29] The calendars were extremely important because they were based upon thousands of years of environmental observations that led to human rotational grazing systems, enhanced by tools such as reseeding of yam tips and allowing sufficient time for regrowth; seasonal knowledge that contributed to spatio-temporal resource management with fire; indicators of expected or suitable harvest times, or when not to harvest as it was the breeding season for a particular animal; and the fostering of ecosystem engineers and keystone species within the plant and animal world so that ecosystems could largely run themselves under the careful management of the overarching Aboriginal keystone culture.

An important mechanism for preventing environmental overexploitation are food taboos, which exist across almost every society. Reinforced by religious beliefs, they are believed to have originated for functional ecological and medical reasons. Known variously as moieties, totems or skin groups (they were called *couberne* by the Wheelman group at Jerramungup), every Aboriginal person had special animals and plants that they were not allowed to eat or destroy. This was passed on through matrilineal descent.[30,31] Kelly Flugge, an Elder from Bremer Bay east of Albany on Western Australia's south coast, explained to me that not only was a person not allowed to eat the animals or plants in their skin group, but they had the responsibility of looking after it/them and preventing overexploitation by others.[32]

Another common strategy involved enforcing food restriction upon people at different life stages, as explorer Edward John Eyre observed:

> There are many usages in force among the natives respecting the particular kinds of food allowed to be eaten at different ages; restrictions and limitations of many kinds are placed upon both sexes at different stages of life.[33]

These restrictions did not apply to children up to the age of about 10, 'no restrictions are placed upon very young children of either sex' or in old age, 'men and women are allowed to eat anything, and there are very few things that they do not eat'.[33] Hassell also described food restrictions in the Wheelman group determined by age and gender, 'young people were not allowed to eat wild dog or eaglehawk. [If they catch them] they were required to bring them to camp and hand them over to the old people'.[15] Certain meats could only be eaten by men and others only by married couples. The complex institutions of restrictions such as these would have reduced impacts on specific food species. Food taboos ensured a wide resource usage through time and space, thus preventing the overuse of the most palatable species.[34,iii]

Sharing of resources with other species that also used them for food was an important practice. Aboriginal people engaged in fruit harvesting would leave enough for other animals that also ate that fruit. Flugge describes how his grandfather taught him to share berries from the *Mulya* bush with bobtail and bluetongue lizards, by leaving the low fruits for the lizards.[32] There were animal hunting behaviours that left enough breeders to repopulate; fishing that left the big fish for future breeding; not hunting animals during their breeding season; looking after important food and habitat trees to protect them from burning; and maintaining the right plants (grass trees were particularly important) as havens for insect populations and animal food.[35,iv] The importance of conservation reserves and sanctuaries for breeding and nesting are probably underestimated. Dame Mary Gilmore wrote about sanctuaries in New South Wales, describing numerous Aboriginal reserves and sanctuaries for various species in the Murrumbidgee and Wagga Wagga area where she grew up. She contrasted Aboriginal encouragement of species numbers with the settlers' lack of management:

> … when I asked my father why we could not get fish as formerly he said, 'When the blacks went the fish went': meaning that the habit of preserving the wild was destitute in the ordinary white settler. Yet at that time the white population on the rivers was only a fraction of what the black had been.[9]

Completely protecting some Country from hunting and harvesting was a vital strategy. Anthropologist Deborah Bird Rose explains, '[if there is] a Dreaming site that is focused on a nesting or breeding area, and there is a prohibition on hunting in that area, there is effectively a refuge in which the special species, and all the other species who use the area, are safe from human predation'.[36] Sanctuaries were often sacred sites and known as increase sites, in which abstaining from hunting, gathering, fishing and sometimes burning contributed to the overall environment. In terms of eco-mythology, Aboriginal people fashioned their belief systems for a high chance of success, which ensured that sanctuaries were landscape spaces in the best habitat for the animal and plant species concerned. The sanctuaries also ensured the long-term viability of a group's home range, for they allowed biota overflow to rapidly restock outside areas after droughts and other major environmental perturbations. They also provided a source of breeder animals that could be caught and transported to restock depleted areas.[37]

These eco-cultural strategies are well recognised methods in numerous Indigenous societies. They lessened deleterious human impacts in time and space by establishing restraints on what could be harvested, when, where and by whom, and the times to move somewhere

---

iii Gender roles are acknowledged in hunting–foraging societies.
iv Ngadju fire practices were designed to protect various forms of vegetation habitats.

else before the ability of food plants and animals to regenerate was compromised.[38] The vast amount of time that such systems have successfully functioned for a wide range of animal and plant species, each with particular ecosystem requirements, is my reason for placing them at the pinnacle of human ecoagricultural food production and management systems.

Long-term protection of food stored in the landscape's ecosystems was a way of providing for planned future ceremonies and populous gatherings of different groups, as was seasonal abundance of particular foods. For a short time, restraints could be released, and profligate and lavish food use could take place at the large ceremonial communal gatherings such as the Bunya festival described by Ludwig Leichhardt in 1847, 'The whole of these people were on their way to the Bunya Bunya Country, for the purpose of obtaining that very remarkable fruit, the product of the Bunya pine' (*Araucaria bidwellii*). This tree is found in the Bunya mountains and Blackall Range in the Darling Downs region of southern Queensland. It has a two- to three-year cycle of abundant fruiting, producing large numbers of nuts. Until the very early 20th century it was a focal point where Aboriginal people gathered to partake of the delicious nuts (some people were reported to have come from as far as South Australia and Western Australia) and to conduct social and cultural business.[1,39]

The ecosystem management practices of Aboriginal societies were also practised in other traditional societies throughout the world. They seem to have become a universal human strategy developed when landscapes lost their megafaunal ecosystem engineers and many keystone species. Research by ecologist Juha Pykälä shows that, in Europe, long-term traditional animal husbandry (now under threat from agricultural industrialisation), which uses unfertilised semi-natural grassland mosaic patches within forests to graze cattle, partially replaced many human-suppressed important natural processes that ceased after the human overkill of European herbivorous megafauna and cohort species. The traditionally managed pastures require the application of fire, grazing and mowing to maintain open woodlands and encourage disturbance, allowing biodiversity and ecosystem function on the wider scale. As such, they mimic many features of deep past animal landscapes.[40] Similar to Aboriginal SES, it is evident that traditional European SES were managed to create a heterogenous mosaic of animal/plant landscapes.

In Australia, fire was used in a sophisticated manner to create and maintain ecosystems in a mosaic of grasslands and bushlands at different stages of post-burn recovery, suiting a range of plant and animal species. In the Esperance bioregion, grasslands and grassy bushlands were maintained for the animals that lived there, such as kangaroos, wallabies, emus and bush turkeys. These habitats were interspersed with mallee and yate forest that provided habitat for possums and various dense thicket-dependent wallabies and small marsupials. Heathlands yielded yams and rhizomes, banksias, grevilleas and hakeas as habitat for small food mammals, such as pygmy and honey possums, bandicoots and many species of birds, especially in and around the dense paperbark swamps. Other food sources were fish and shellfish obtained from the estuaries and seashores.

The discontinuation of Aboriginal culture as the keystone cultural force was and is an important cause of the cascading extinctions of small mammals, especially in arid areas.[41] For example, the burrowing bettong or *mitika* was once widespread throughout arid to semi-arid eastern Australia, Central and Western Australia. It is considered a keystone species which maintained grasslands in concert with Aboriginal people. Roaming several kilometres from their extensive underground warrens, bettongs fed on and suppressed the acacia seedlings regenerated by Aboriginal firing, and so maintained grassy mosaics. Since their disappearance, dense stands of acacia scrub have regenerated and now dominate much of the landscape.[42] Similar examples of past co-evolved animal landscapes include the visible remnants of wombat-maintained clearings, especially around granite domes in the eastern woodlands of the Esperance bioregion.

Ecoagriculture with fire and other tools

Human relationships with fire extend back to *Homo erectus* at least 1 million years BP. The relatively small teeth of *Homo erectus*, compared to those of other hominids, provide evolutionary evidence of food cooked with fire. Hominids' routine use of fire to burn landscapes after leaving Africa was likely to have been well established by approximately 400 000 years ago.[43] Aboriginal extension and maintenance of habitat through deliberate and sophisticated use of fire was first proposed by Norman Tindale in the 1950s, supported by Tasmanian anthropologist Rhys Jones in the 1960s and confirmed by Sylvia Hallam's research in south-western Western Australia during the 1970s.[44,45,46] When Rhys Jones coined the term 'firestick farming' to describe Aboriginal land management, he was deliberately provocative in applying the word 'farming' to a people who had been believed to have no farming or agricultural abilities. As ecoagriculturists with a unique 'farming' methodology, fire was the most widely used tool.

Fire use in the Esperance bioregion was first observed in 1791, when French explorers aboard the frigates *L'Espérance* and *La Recherche* realised the presence of human habitation when they saw smoke. Botanist Jacques-Julien de Labilliardière noted on 9 and 10 December:

> We had not yet seen the least indication of inhabitants ... and accordingly, the smoke of two fires, which they had kindled, convinced us of their presence ... Other fires lighted along the coast, sent up large columns of smoke ...[47]

The first smoke seen was probably from cooking fires, the second seems to indicate that some sort of management burn was in progress. Labilliardière interpreted the fires as the locals signalling to the French, which reflects the narrow understanding of the uses of fire in European cultural beliefs. Aboriginal people used fire in their landscapes for a far wider range of purposes than could be imagined by Europeans.

Later, Claude Riche, one of the naturalists on the expedition, was lost for several days around Pink Lake near the present town of Esperance. During this time, he observed Aboriginal people at their fires:

> They put to fire a field covered with bushes and propagate the flames until everything has been consumed ... I saw, the day after, several natives who remained from morning till night poking their fires.

Riche further asked:

> But what, perhaps, is the cause which obliges them, in a hot climate, to make considerable fire all day long? This usage is widespread in a great part of New Holland, because we have it in practise also in Van Diemen's Land.[47]

He described walking through a landscape patchy with open areas and bush, where fire had been used to completely remove bushes from a 'field'. The valley bottoms were covered in marshes and 'the hillocks up to certain height shaded by large trees, and high grass ... A very numerous species of plants and a great variety of birds rendered this more vivacious'.[47] Unwittingly, he was seeing answers to his own question – a landscape of vegetation mosaics being deliberately managed with fire to encourage plant and animal biodiversity.

In 1848, after traversing vast tracts of seemingly uninhabited northern mallee lands in the Esperance bioregion, John Septimus Roe described the settled appearance of the coast west of Point Malcolm to Esperance Bay. He noted that native fires caused large amounts of smoke and that Aboriginal people, kangaroos and emus occurred together along the coast, among 'grass trees, xamia and yeit trees' with a grassy understorey. Further on he described 'large smokes' from fire west of the granite hill that he named Mt Merivale.[17] John Forrest's journal, written around 30 years later before many of the changes arising from European occupation, described similar scenes. East of Esperance, he described how beautiful the country was, with patches of grasslands and numerous brackish streams interspersed with heathlands.[48]

After meeting a party of 'natives' north of Albany in November 1840, J.L. Stokes from HMS *Beagle* described with awe:

> The dexterity with which they manage so proverbially a dangerous agent as fire is indeed astonishing. Those to whom this duty is especially entrusted, and who guide or stop the running flame, are armed with large green boughs, which if it moves in a wrong direction, they beat it out ...[49]

Fire was so important to Aboriginal SES that a word for fire in south-western Australian Nyungar language, *karl*, not only describes fire but also the immediate or nuclear family.[50,51] Fire provided warmth and light and heat for cooking, as well as the foundation for important ceremonies and smoke for cleansing the spirit. There is evidence in explorer and settler journals that Nyungar burning in the south-west and the south coast was usually carried out from December to March. The frequency of these fires meant that they were of low intensity – nothing like the wildfires that have become frequent today.[52]

There were many and varied rules associated with the use of fire in Australian landscapes that produced resilient management regimes adaptive to the cycles of nature and prevailing climate conditions. Frequent burns promoted specific plant and animal communities, less frequent burns affected other communities; hot burns, cool burns, burns of differing extent and at different times of the year – these are only a few of the myriad fine-grained permutations that existed. Sylvia Hallam describes how firing by men and women differed in range, type of country and intensity since each group was obtaining food from different types of habitat and so managing for different animal and plant species and species suites.[46] Research by Western Australian Conservation and Land Management fire expert David

Ward and others has shown that a fine-grained mosaic can only continue to be a stable landscape feature if patches are burnt when they are only just able to carry a fire.[52] It was common throughout Australia for Aboriginal people to declare that their burning practices were 'cleaning up Country', so preventing destructive high-intensity wild fires.

The role of specific fire regimes in maintaining habitat for ground-dwelling mammals in Western Australia was vital and has been investigated by zoologists G. Friend and A. Wayne. They found that a reasonably intense fire will initially reduce the woylie population but within four to five years the population density will recover as food supplies increase, allowing breeding and migration of young animals to unburnt surrounding habitats. Intense fire is needed to create the tammar wallaby habitat of dense scrub thickets that provide shelter from predators, but they need adjacent grassy, shrubby areas in which to graze. Fire was used to aid the hollowing of fallen logs which small marsupials such as the numbat (*Myrmecobius fasciatus*) need for shelter. Other fire regimes, as well as physical interventions such as beating out flames, create buffers around habitat that needs protection from burning, such as the deep litter layers with their associated insects and other arthropods, vital as food for animals such as the southern brown bandicoot (*quenda, Isoodon obesulus*) and yellow-footed antechinus (*mardo, Antechinus flavipes*).[53]

Fire was also very important in Ngadju Country. On the coastal lands, fire was used over the summer months but inland Country was not fired during the hot season (*Nganji*, November to March). Research and oral histories obtained by Susan Prober and partners describe outcome-specific fire regimes in the mallee woodlands. Most fires were cool, to protect the many fire-sensitive assets including old growth habitat trees (*ngarta*), such as salmon gums (*marrlinja*) and gimlets (*jooderee*) with habitat and living spaces for insects, lizards, birds and small mammals. Cool fires also allowed germination of important food plants (but not too much, as happens with a hot fire) such as the tuber-producing twining fringe lily (*junkajunka, Thysonatus patersonii* or *T. mangliesianus*) and apple berry or Ngadju bush (*ngadjun, Billaridiera lehmanniana*) that protect water rockholes and springs that can be damaged by fire. Fire was also used to create and enlarge water-storing *gnamma* holes in granite outcrops, and for logs and log hollows that are animal homes (the more hollows, the more animals), water trees, grass trees and habitat trees for *bardi* grubs such as jam (*murren, Acacia acuminata*) and fruit trees (quandong, sandalwood, chuck), various medicine bushes and mallee fowl. Fire is also important to maintain and regenerate the grasslands that are good for kangaroos. Dry grass is best to fatten the inland red kangaroos (*marlu, Macropus rufus*) and green grass fattens grey kangaroos (*kulpirr, M. fuliginosus*) which live at or closer to the coast. Grassy areas need to be burnt more often, sometimes every year, to keep producing kangaroo and wallaby food as well as to keep the country open for camping and hunting.[35] The types of fire were tailored to the requirements of different species, which could have overlapping needs.

For some situations, fire use was not appropriate or necessary to manage habitat. In her research in the coastal Esperance bioregion, Moya Smith found evidence of micro-habitat creation in the form of lizard traps (refuge habitat) made by people on the five rock outcrop sites she investigated. These lizard shelters were made from small naturally exfoliated slabs of granite propped on smaller stones with an opening on one side. Smith suggests the term 'trap' is a misnomer, as they were made to increase refugia, primarily for king skinks (*wandie*, *Egernia kingii*) which were an important food.[54] If lizard distribution and abundance were limited not by food availability but by the amount of shelters available, increasing shelters could increase the lizard population as well as its hunting predictability.

Modern research is providing increasing evidence about the extent and depth of landscape management systems applied by Aboriginal peoples. The landscape spatial pattern supplying the needs of the widest range and abundance of animals is a mosaic of interconnected bush, open grassland with scattered trees and areas of large trees and very dense bush creating the many and varied edges where habitat types meet. Called the 'edge effect', these patterns maximise the number of habitat edges and therefore resource abundance and diversity, and promote the greatest biodiversity, abundance and distribution of animals.[55] Research in Victoria confirms that significantly more individual birds and bird species are found at the edges of forest and grassland than in the deep forest interior.[56] Aboriginal knowledge about burning or clearing to create the edge effect long pre-empts the processes that professional foresters and conservation managers have only rediscovered after years of research.[57]

Ecoagriculture

Aboriginal people were supreme ecofarmers and observers of non-human nature because their livelihoods both as individuals and as a group depended directly upon ecological knowledge and applying it to everyday intellectual thought. Unfortunately, this cultural characteristic and its implementation for food production and consumption is an addendum in 'civilised' societies ruled by short-term economic and social cycles. Food production sustainability is not automatically viewed as pre-emptive and a long-term fundamental necessity. Modern attempts at conservation landscape management with fire and poisons would seem crude, primitive and probably wasteful to the past Aboriginal experts, for their motivations are not food-oriented. The divesting of biodiverse food resources that provided adaptability to Australia's fluctuating environmental conditions, often climate-related, and replacing them with a minimum of agricultural species and systems needing predictable, stable climatic requirements for production would have also seemed foolish to the past ecoagriculturists. Unlike today, land was not being farmed to produce only a few species of animal and plant, for economic reasons, with sustainability of secondary importance. Nor was it being managed for esoteric, aesthetic and ideological conservation purposes in non-farmed areas. The powerful motivating force was to produce abundant life forms as immediate food and storage systems of living biomass, and the ecosystem support systems for this food and other consumables in perpetuity. The results were the beautiful, diverse and productive ecological systems that were constantly commented upon by the first European observers, and the strong, healthy people who lived within them.

Endnotes

1. Pascoe B (2014) *Dark Emu Black Seeds: Agriculture or Accident?* Magabala Books, Broome.
2. Groom L (2012) *A Steady Hand: Governor Hunter & His First Fleet Sketchbook.* National Library of Australia Publishing, Canberra.

3. Mitchell TL (1838) *Three Expeditions into the Interior of Eastern Australia*. Vol.1. T. and W. Boone, London.
4. Anderson EN (1996) *Ecologies of the Heart*. Oxford University Press, London.
5. Flannery T (1999) *Throwim Away Leg: An Adventure*. Text Publishing, Melbourne.
6. Diamond J (2005) *Collapse: How Societies Choose to Fail or Succeed*. Penguin, Melbourne.
7. Penn DJ (2003) The evolutionary roots of our environmental problems: toward a Darwinian ecology. *Quarterly Review of Biology* **78**(3), 275–301. doi:10.1086/377051
8. Sveiby K, Skuthorpe T (2006) *Treading Lightly: The Hidden Wisdom of the World's Oldest People*. Allen and Unwin, Sydney.
9. Gilmore M (1934) *Old Days: Old Ways – A Book of Recollections. With Notes on the Life of Dame Mary Gilmore* [by Barrie Ovenden]. Angus and Robertson, Sydney. Reprint 1986.
10. Smith EA, Wishnie M (2000) Conservation and subsistence in small-scale societies. *Annual Review of Anthropology* **29**, 493–524. doi:10.1146/annurev.anthro.29.1.493
11. Codding BF, Bird RB, Kauhanen PG, Bird DW (2014) Conservation or co-evolution? Intermediate levels of Aboriginal burning and hunting have positive effects on kangaroo populations in Western Australia. *Human Ecology* **42**(5), 659–669. doi:10.1007/s10745-014-9682-4
12. Cullen LE, Grierson PF (2009) Multi-decadal scale variability in autumn–winter rainfall in south-western Australia since 1655 AD as reconstructed from tree rings of *Callitris columellaris*. *Climate Dynamics* **33**, 433–444. doi:10.1007/s00382-008-0457-8
13. Panter-Brick C, Layton RH, Rowley-Conwy P (Eds) (2001) *Hunter-gatherers: An Interdisciplinary Perspective*. Vol. 13. Cambridge University Press, Cambridge.
14. Winterhalder B (2001) The behavioural ecology of hunter-gatherers. In *Hunter-gatherers: An Interdisciplinary Perspective*. Vol. 13. (Eds C Panter-Brick, RH Layton, P Rowley-Conwy) pp. 12–14. Cambridge University Press, Cambridge.
15. Hassell E (1936) *My Dusky Friends*. Republished by C.W. Hassell, Perth, 1975.
16. Kirschenman F (2010) *Cultivating an Ecological Conscience: Essays from a Farmer Philosopher*. Kentucky University Press, Lexington.
17. Roe JS (1849) Report on an expedition to the south-eastward of Perth, in Western Australia, between the months of September 1848, and February 1849 under the Surveyor- General Mr. John Septimus Roe. *Journal of the Royal Geographical Society of London* **XXII**, London.
18. Whaley Eager P (2017) *Global Population Policy: From Population Control to Reproductive Rights*. Routledge, New York.
19. Bignell M (1971) *First the Spring: A History of the Shire of Kojonup, Western Australia*. Shire of Kojonup, Western Australia.
20. Berry W (1997) *The Unsettling of America*. Sierra Club Books, San Francisco.
21. Cowlishaw G (1981) Determinants of fertility among Australian Aborigines. *Mankind* **13**(1), 42–45; Love JRB (1936) *Kimberley People: Stone Age Bushmen of Today*. David M. Welch [reprinted 2009], Australian Aboriginal Culture Series No. 6.
22. Webb LJ (1969) The use of plant medicines and poisons by Australian Aborigines. *Mankind* **7**(2), 137–146.
23. Lobdell JE (1975) Considerations on ritual subincision practices. *Journal of Sex Research* **11**(1), 16–24. doi:10.1080/00224497509550873
24. Reynolds H (2006) *The Other Side of the Frontier: Aboriginal Resistance to the European Invasion of Australia*. University of NSW Press, Sydney.
25. Nind IS (1831) Description of the natives of King George's Sound (Swan River Colony) and adjoining country. *Journal of the Royal Geographical Society of London* **1**, 21–51. doi:10.2307/1797657
26. Green N (1984) *Broken Spear: Aboriginals and Europeans in the Southwest of Australia*. Focus Education Services, Perth.
27. Mitchell MB (2016) The Esperance Nyungars, at the frontier: an archaeological investigation of mobility, aggregation and identity in late-Holocene Aboriginal society, Western Australia. PhD thesis, Australian National University, Canberra.
28. Gara T, Cane S (1988) *Environmental, Anthropological and Archaeological Background to the Nullarbor Plains*. Report to Heritage Branch, Department of Environment and Planning, South Australia, by ANUTECH Pty Ltd. Australian National University, Canberra.

29. Prober SM, O'Connor MH, Walsh FJ (2011) Australian Aboriginal peoples' seasonal knowledge: a potential basis for shared understanding in environmental management. *Ecology and Society* **16**(2), 12. doi:10.5751/ES-04023-160212
30. Meyer-Rochow VB (2009) Food taboos: their origins and purposes. *Journal of Ethnobiology and Ethnomedicine* **5**, 18 doi:10.1186/1746-4269-5-18.
31. Hassell E, Davison DS (1936) Notes on the ethnology of the Wheelman tribe of south-western Australia. *Anthropos* **31**(5/6), 679–711.
32. Flugge L (2014) Nyungar Nation Elder and Aboriginal Projects Foods Manager. Interviewed by Nicole Chalmer, Department of Agriculture and Food, Perth.
33. Eyre EJ (1845) *An Account of the Manners and Customs of the Aborigines and the State of their Relations with Europeans*. T. and W. Boone, London.
34. Brightman R (1996) The sexual division of foraging labour: biology, taboo and gender politics. *Comparative Studies in Society and History* **38**(4), 687–729. doi:10.1017/S0010417500020508
35. Prober SM, Yuen E, O'Connor M, Schultz L (2013) *Ngadju Kala: Ngadju Fire Knowledge and Contemporary Fire Management in the Great Western Woodlands*. CSIRO Ecosystem Sciences, Perth.
36. Rose DB (1996) *Nourishing Terrains: Australian Aboriginal Views of Landscape and Wilderness*. Australian Heritage Commission, Canberra.
37. Newsome AE (1980) The eco-mythology of the red kangaroo in Central Australia. *Mankind* **12**(4), 332–333.
38. Folke C, Colding J, Berkes F (2003) Synthesis: building resilience and adaptive capacity in social-ecological systems. In *Navigating Social-ecological Systems: Building Resilience for Complexity and Change*. (Eds F Berkes, J Colding, C Folke) pp. 352–387. Cambridge University Press, Cambridge.
39. Gerritson R (2008) *Australia and the Origins of Agriculture*. Archaeopress, London.
40. Pykälä J (2000) Mitigating human effects on European biodiversity through traditional animal husbandry. *Conservation Biology* **14**(3), 705–712. doi:10.1046/j.1523-1739.2000.99119.x
41. Morton SR (1994) European settlement and the mammals of arid Australia. In *Australian Environmental History: Essays and Cases*. (Ed. S Dovers) pp. 158–159. Oxford University Press, Melbourne.
42. Noble JC, Gillen J, Jacobson G, Low WA, Miller C, the Mutitjulu Community (2001) The potential for degradation of landscape function and cultural values following the extinction of mitika (*Bettongia lesueur*) in Central Australia. In *Land Degradation*. (Ed. J Conacher) pp. 71–89. Kluwer Academic, Dordrecht.
43. Doughty C (2013) Pre-industrial human impacts on global and regional environment. *Annual Review of Environment and Resources* **38**, 503–527.
44. Tindale NB (1959) Ecology of primitive Aboriginal man in Australia. In *Biogeography and Ecology in Australia*. (Ed. A Keast *et al*.) pp. 36–51. Springer, Dordrecht.
45. Jones R (1969) Firestick farming. *Australian Natural History* **16**(7), 224–228.
46. Hallam S (1979) *Fire and Hearth: A Study of Aboriginal Usage and European Usurpation in South-western Australia*. Advocate Press, Melbourne.
47. De La Billardiere J-J (1808) An account of the expedition. In *The Voyage of D'Entrecasteaux*. (Ed. M De Rossel) Imperial Press, London.
48. Forrest J (1875) Last Day in Esperance Bay… Extracts from *Explorations in Australia*. Cambridge University Press, Cambridge.
49. Stokes JL (1886) *Discoveries in Australia; With an Account of the Coasts and Rivers Explored and Surveyed during the Voyage of H.M.S. Beagle in the years 1837–43*. T. and W. Boone, London.
50. Whitehurst R (1997) *Noongar Dictionary: Noongar to English; English to Noongar*. Noongar Language and Culture Centre, Perth.
51. Burrows N (2009) A fire for all reasons. In *Landscope, Fire the Force of Life*. Special edition, Vol. 1. Department of Environment and Conservation, Perth.
52. Ward D (2015) Fire ecologist, Department of Environment and Conservation (retired). Interviewed by Nicole Chalmer, Perth, 2 June.
53. Friend G, Wayne A (2003) Relationships between mammals and fire in south-west Western Australian ecosystems: what we know and what we need to know. In *Fire in Ecosystems of South-West Western Australia: Impacts and Management*. (Eds I Abbott, N Burrows) pp. 365–367. Backhuys Publishers, Leiden.

54. Smith M (1993) Recherché a l'Esperance: a prehistory of the Esperance region of south-western Australia. PhD thesis, University of Western Australia, Perth.
55. Ries L, Fletcher JR, Battin RJ, Sisk TD (2004) Ecological responses to habitat edges: mechanisms, models, and variability explained. *Annual Review of Ecology Evolution and Systematics* **35**, 491–522. doi:10.1146/annurev.ecolsys.35.112202.130148
56. Berry L (2001) Edge effects on the distribution and abundance of birds in a southern Victorian forest. *Wildlife Research* **28**(3), 239–245. doi:10.1071/WR00057
57. Meilleur BA (1994) In search of 'keystone societies'. In *Eating on the Wildside: The Pharmacologic, Ecologic, and Social Implications of Using Non-cultigens*. (Ed. NL Etkin) pp. 259–279. University of Arizona Press, Tucson.

5
Healthy ecosystems, healthy food, healthy people

> While good science should be ruled by scepticism and self-doubt, the field of nutrition has instead been shaped by passions verging on zealotry.[1]

The approach to food in our society has become an obsession with reductive microanalysis about all that is good or bad for us to eat. An overload of bad science promoting mistaken past beliefs about hunter-gatherer diets being comprised of 75% plant foods and only 25% animal-based foods has contributed to the doctrine that too much red meat is bad, and a majority plant-based is good.[2] This was reinforced by decades of faulty research conducted by the 20th century's most influential nutrition expert from the United States, Ancel Keys and cohort, who attributed the rise of heart disease in the US post World War II to saturated animal fats in the diet.[3] The Australian nutritional food pyramid still contains many aspects of Keys' dogma, and nutritionally incorrect interpretations of hunter-gatherer diets which placed fruit and vegetables as the majority of their diet, followed by grains, then lean low-fat animal products and legumes, with plant-based oils and fats at the top.[4]

Many of these dietary recommendations have features in common with those of the early settled agricultural societies in the Americas (without animal domesticates) where grains such as maize, along with vegetable foods, became the dominant dietary components as wild game animals were hunted out. The prevalence of iron deficiency-caused anaemia was dramatically higher in such societies than in those of their hunting and gathering ancestors, who had far greater varieties and amounts of meat and animal fats in their diet. Further indicators of poor nutrition were bone lesions caused by infection, which grew four-fold in early agricultural Americans, dramatically increasing the prevalence of osteoarthritis and significantly shortening lifespans.[5]

Humans are omnivores adapted to eating animal protein and fats and, less efficiently, lower-cellulose higher-nutrient plant storage or reproductive foods (tubers, nuts, seeds, fruits and some leaves). Our relatively short, unspecialised digestive systems (though not as short as those of specialised predators such as cats and native cats) are not designed for large-scale plant fermentation and digestion. True herbivores have specialised fermentation 'vats' as part of their digestive systems, as either a rumen or greatly enlarged caecum. The ruminant herbivores (such as cattle, sheep, goats, deer and their relatives) have an enormous vat – the rumen, which is a second stomach designed for breaking down indigestible plant cellulose with the help of specialised bacteria, protozoa and fungi. The caecum digesting herbivores (such as rhinos, horses and rabbits) have a large specialised caecum that digests cellulose with the aid of specialised microbes. These true herbivores need to spend at least eight to nine hours per day grazing to get sufficient plant material for the fermentation microbes to extract the macro and micronutrients needed. Much of their protein comes from digesting dead fermentation microorganisms.

In terms of return on effort both in catching and digesting, animal-based foods are superior to plant-based foods because of their far higher nutritional benefits. As nutritional anthropologist Neil Mann explains, the evolution of meat- and fat-based diets in humans is based on their superior nutrient and micronutrient return for energy output. Furthermore, our evolutionary requirements are reflected in various digestive inabilities: humans cannot synthesise the amount of taurine needed (an essential amino acid obtained easily from meat) from plants, and only animal-based foods ensure adequate levels; we lack the ability to chain elongate plant fatty acids; further evidence of meat as normal in human diets are the ancient lineages of co-evolved parasites related to dietary meat animals; and animal foods contain proteins and a range of fats, including saturated fats which contain vitamins and fatty acids (vital for various cell functions) which humans are unable to synthesise from plants. Our inability to synthesise vitamin C, shared with other primates and guinea-pigs, means that plants or animal parts with this micronutrient are essential in the human diet.[6]

Aboriginal ecoagricultural systems provided a diet dominated by animal protein and fats, comprising at least 75% of the dietary intake. Animals from insects to reptiles, mammals and birds that had seasonally high amounts of fat were the most prized, as wild animals are generally lean and fats and oils are not easy to obtain. Another feature of this diet was its wide diversity of plant foods of higher nutrient density than most modern plant foods that have been selectively bred for sweetness and appearance rather than nutrition. This food diversity was observed by Ethel Hassell at Jarramongup who wrote that 'practically everything edible was eaten' – water and wetland plants and animals, large and small marsupials, birds, reptiles, amphibians and insects such as the fat-rich *bardi* grubs (the wood-eating larvae of various moths and beetles).[7] Lerp, the sugary candy coating of the sap-sucking larvae of the insect family Psyllidae, was harvested, along with many plant fruits, seeds, leaves, tubers, roots and fungi species.[8,9]

Throughout Australia, ecosystems always contained more potential food species than were actually eaten, as the effort required to harvest exceeded their nutritional value and perhaps some didn't taste good. Ngadju Elder Sonny Graham explained to me that though there were many kangaroos living around Israelite Bay they were not considered worth eating because they did not taste good – he attributed this to a certain type of plant they ate in this location.[10] Ngadju woman Dorothy Dimer dismissed the abundant and easily caught bearded dragon lizard (*Pogona minor*) as a food choice because of its lack of meat and its ordinary taste.[11] With such abundant food available, despite various cultural restrictions people could choose what to include in their diet. In 1841 explorer George Grey observed that for Aboriginal people there were plant foods in season in every period of the year, and 'the natives regulate their visits to the different districts accordingly'.[12]

Traditional Aboriginal diets were among the healthiest in the world. Like many indigenous peoples they had exceptional teeth, vision and hearing, and were nearly disease-free and often long-lived.[12,13] Numerous historical accounts describe Aboriginal people as extremely well fed, healthy and well built. As George Grey emphasised in 1841, 'the mistake very commonly made with regard to the natives of Australia is to imagine that they have small means of subsistence or are at times greatly pressed for want of food'.[12] In comparison to the average English person in the 19th century, Aboriginal people were better fed, healthier and had a longer lifespan. Grey deduced that most of the people he had interactions with lived well past 70. About the old men, he noted that 'they seldom appear to suffer much from the infirmities and diseases to which the aged are generally subject amongst us'.[12] This is supported by Ethel Hassell, who remarked how 'disease seemed rare amongst them'. Hassell also recounted a trip to Esperance Bay by her husband Albert in 1861 when he was 19, in which he spent time with a portion (200 people) of the Kar Kar group whose boundary extended to Esperance Bay. He described them as 'a very fine-looking lot of natives, all tall and well made'. There was one family he particularly remembered, 'six boys and not one under six feet'.[7]

Until recently, researchers believed that traditional Aboriginal diets were dominated by plant foods, in the often-suggested ratio of 75% or greater plants to 25% or less animal protein. A ratio close to this was determined by Richard and Pamela Smith from a long history of research in the Western desert.[14] This research is supported by many older studies on this type of harsh environment, but should not be extrapolated to more benign bioregions where physiological research, such as reported by J.B. Miller and Stephen Colagiuri, concludes that there are high levels of genetically determined insulin resistance in Aboriginal populations. This provides strong evidence for a diet comprising at least 75% animal protein.[15]

Anthropologist Sara Meagher has documented a vast amount of early explorer and settler observations about Aboriginal food in Western Australia. This, along with other historical and social research, supports the importance of meat in the diet. These resources about early contact provide evidence for the meat dominance in diets as they describe the food resources of south-west and southern coastal Aboriginal people – animal foods were the dominant and preferred diet, with plant foods a supplement.[6,16] It seems, then, that in most of Australia diets were dominated by animal-based protein with vegetables, grains and fruits a supplement, with desert diets an exception to the majority or perhaps reflecting the early impacts of pastoralism on desert fauna.[13,17,18] Aboriginal Elder Sonny Graham confirmed this when he explained that eating mostly animal foods was always the preferred option.[10]

For Aboriginal people throughout Australia, the most favoured foods were safe organ meats and the fats attached to them. Emus, with their high proportion of soft fat, were notably esteemed as food. Their fat was also massaged into the skin and used to ease joint pains in older people. Fat was highly valued. Kangaroos in the northern Esperance bioregion laid down fat during the early dry season and were best to eat then, according to Ngadju people.[19] Throughout Australia, animals were preferred when they were fattest because a high-protein diet without enough fatty acids may lead to illness or even death. Ethnologist Heather McDonald describes a Jaru woman from the east Kimberleys who explained:

> Fat makes us *munda yura* [good belly]. We get the fat from the kangaroo guts, peel the fat off the guts, grill it and eat it. We grill the kangaroo fat and fill ourselves up. Liver fat is the best tucker. Cook the liver fat and eat it.[20]

The fat ratio of omega 6 and 3 oils at close to 2:1 in wild grass-fed animals (and oily fish such as wild salmon) is close to the ideal. In contrast, grain-fed animals can have a ratio of up to 20:1. There are several studies suggesting that meat from grass-fed herbivores elevates the precursors for vitamin A and E, as well as cancer-fighting antioxidants such as glutathione (GT) and superoxide dismutase (SOD), arguably the body's most crucial antioxidant.[14,15] Wild animals also have high levels of the healthy linoleic acid and a healthy ratio between saturated, monounsaturated and polyunsaturated fats. Oils from plants were of lesser importance in the diet, as fewer plant species concentrate fats and oils compared to animal species. Quandong, sandalwood kernels and macadamia nuts have abundant oil, and some native grasses have seeds relatively high in protein and linoleic acid.[20]

Aboriginal groups exhibited a marked division of labour in obtaining food, with men generally hunters of larger animals, such as western and eastern grey kangaroos and emus. The women (and children) hunted smaller animals and were the main collectors and managers of plant foods.[21] Men as larger animal managers were concerned with how best to manage kangaroos, emus and other larger prey, using burning regimes to encourage grass and feed for them. However, they also knew how to find plants and smaller animals when necessary. Women were the supreme botanists, horticultural experts and small animal managers concerned with the procurement of various plant products as well as smaller vertebrates and invertebrates. Firing by men and women varied in range, type of country and intensity, as each sex was managing different types of habitats for different species and species suites. The sexes cooperated when needed to manage fire regimes suitable for different habitat requirements and would hunt cooperatively as well.[22] Tom Dimer from Nanambinia Station in the eastern Esperance bioregion woodlands described in a 1990 interview for the Western Australian State Historical Archives, how men and women would construct a bush fence raceway in tammar wallaby habitat, judiciously using fire and pet dingoes to drive the wallabies down it, killing only those they wished to eat and letting the others go free to breed for the future.[23]

The gendered division of labour was reflected in the type of implements used. Spears, axes, clubs, throwing sticks and boomerangs were owned and used by men, while the digging stick was the main implement favoured by women. It was extensively used and could be a weapon as well as a digging implement and planting tool. At times, all members of a group would work together to catch food. Nets were used to catch fish, birds and even animals as large as kangaroos and emus. Land management, to best ensure the full range of continuous plentiful food resources, required input, discussion and cooperation between men and women. For example, determining what to burn, when, where, how intensely and how often would be necessary for a range of habitats for animal and plant species – not just kangaroos and emus. In the Esperance bioregion, given the importance of animal products in Ngadju and Nyungar diets, the focus of the ecologies of landscape management would have been what was best for animal production.

Animals

In 1993, Moya Smith completed her doctoral thesis about Aboriginal foods and lifeways based on evidence found along the south-eastern coast of Western Australia in the Esperance bioregion. Important information was found in and around a cave at Cheetup Hill, 65 km east of Esperance town and situated in the Cape le Grand National Park, immediately behind the farm belonging to myself and my husband.[24] Her archaeological excavations provided evidence of important foods. An ancient pit was uncovered containing macrozamia seeds, and radio-carbon dating of charcoal in the overlying deposit showed an age of over 13 000 years. This proved for the first time that Aboriginal people had known how to process the toxins from macrozamia and thus make them safe for eating, for far longer than previously thought. Excavations at another cave at Barndi (near the ocean at Duke of Orleans Bay) provided bone remains that indicate the eating of animals sourced from granite outcrops, heathlands, woodlands and ocean during the last 2000 years. There were several species of predatory marsupial mice such as *Antechinus,* white tailed dunnart (*Sminthopsis granulipes*), slender tailed dunnart (*S. murina*) and the omnivorous southern brown bandicoot. Grey kangaroo and wallaby species included western brushtail (or black gloved wallaby) and woylies were abundant, along with several native mice such as the heath mouse (*Pseudomys shortridgei*), ash grey mouse (*P. albocinereus*) and western mouse (*P. occidentalis*). Remains of fish, crustaceans, lizards and frogs were also found.[24]

The sites had remains of small, very fast-breeding mammals such as the southern bush rat (*Rattus fuscipes*) – a species documented as exploited by Aboriginal people in the south-west – the honey possum (*Tarsipes rostratus*) and western pygmy possum (*Cercartetus concinnus*), which lives in the mallee as well as the sandplain of the Esperance bioregion.[24] Local Aboriginal Elder Sonny Graham, with whom I discussed food resources, related that 'honey possums tasted like honey' and were a favourite snack during his childhood because of their delicious sweet taste.[10]

In discussions with 93-year-old Edward Hannett, he described foods still available to Aboriginal people in the Esperance bioregion during the early 20th century.[25,i] He was born east of Esperance in 1914 and recalled small mammals that were present when he was a boy, the woylie (*wal* or *waldhoo*) and burrowing bettong (*burdie*). He had a pet black gloved wallaby, who would hop onto the verandah to sleep, and remembered that ringtail possums (*waarmp* or *warder, Pseudocheirus peregrinus occidentalis*) were common in the yate swamps.[25] Ringtail possums build nests (called dreys) in trees, so do not rely on tree hollows for shelter. Tree hollows are important to brushtail possums (*Trichosurus vulpecula*), though they can also use rockholes and deep overhangs when hollows are unavailable. Brushtail possums are well documented in the south-west of Western Australia and throughout eastern Australia as an important staple meat source but appeared to have a patchy distribution in much of the Esperance bioregion, so would have been a localised important food animal.[16,26] At Thomas River Station, with many large yate trees, they were abundant enough to be later hunted for fur by settlers, so would have been important to the local people. In the late 1980s, Karl Dimer of Nanambinia Station south of Balladonia described how he had eaten possum from 'possum country' in the woodlands where larger trees had breeding hollows.[27] Isaac Nind, surgeon appointed to the King George Sound military garrison (Albany) from 1826 to 1829, assiduously recorded Nyungar life. He described how dingoes were used to assist in hunting:

---

i   Mr Hannett identified mammals using the following reference: Strahan R (Ed.) (1991) *The Australian Museum Complete Book of Australian Mammals*. Angus and Robertson, Sydney.

> In the chase the hunters are assisted by dogs ... They are particularly useful in catching bandicoots, the small brush kangaroo and the opossum but for the emu and kangaroo they are not sufficiently fleet. The owner of a dog ... is entitled to an extra portion of the game killed [and is] lent out upon consideration of the owner receiving a share ...[28]

Dingoes were also eaten. Explorer George Grey noted that 'a dog is baked whole in the same manner as a kangaroo'.[12] Hassell and Nind both described how 'only the old people were allowed to eat them'.[7,28] It could be assumed that in the past thylacines and Tasmanian devils were also eaten, as Aboriginal people were the apex (top) predator. The eating or killing of lower-rung predator species (mesopredators) by an apex predator reduces their competitive impacts through fear of death. This changes their behaviours, to avoid perceived risky areas, and has the potential to reduce exploitation competition as well as prevent overhunting of prey populations.[29,30,ii] This control (but not elimination) of dingoes as the only large non-human predator left in Australian ecosystems would have contributed to the biodiversity and abundance of prey species constantly noted by early European explorers and settlers.

Many bird and reptile species were eaten, with reptiles probably the major source of animal protein and fat in arid regions. Throughout Australia in the higher-rainfall regions around the coast and along rivers systems, birds, their eggs and young – especially parrot species – as well as ducks, pigeons, swans and hawks were particularly important food in spring. Mallee fowl and their eggs are attributed by ethnographer Daisy Bates as being of enormous importance to all groups that lived in the more arid mallee country in Western Australia and South Australia. Her point about abundance is well made:

> Mallee hens' eggs at certain periods of the year were to be gathered in millions in the swamps of the far south – west, and then the people of the Ngau, the mallee hen totem, would invite neighbours far and near to a feast extending for many weeks, the light brown egg with their reddy–yellow yolks providing a meal for a million.[31]

With mallee fowl (*gnow*) now rare enough in Australia that it has been assigned the status of 'threatened', it seems almost unbelievable that they could have provided the abundant food described by Bates. However, considering that their prime mallee habitat was one of the most widely destroyed by clearing for wheatlands from the early 20th century, their endangered status and dramatic decline is not surprising. Without major revegetation programs, they will only be able to hang on in the rare bushland habitat patches that escaped clearing.

Parrot and cockatoo species were also important foods. The purple crowned lorikeet called *kinga* (*Glossopsitta porphyrocephala*) was a special delicacy as, like the honey possum, it tasted of the nectar it fed upon.[10] Unfortunately, several other Western Australian species such as Carnaby's cockatoo (*Calyptorhynchus latirostris*) that relied on tree hollows in the mallee woodlands for nesting and once numbered in their millions, are now in a similar or worse position than mallee fowl.

---

ii Ritchie and Johnson show that prey abundance and biodiversity can be increased when an apex predator controls the mesopredators.

Reptiles were a very important meat and fat resource in most of Australia, as they are widespread and common. The same species or subspecies of reptiles are found along the southern coasts of Australia, therefore Nind's descriptions of those used at Albany are relevant. He describes two of the largest species and several smaller but common species.[28] The largest – *munnaar* to Nyungar and *kalun* to Ngadju (the southern heath monitor, *Varanus rosenbergi*) – can reach 1.3 m in length. It is replaced in the mallee by *kalga* to Ngadju (Gould's sand monitor, *V. gouldii gouldii*), which was described by Dorothy Dimer as a delicacy because of its delicious fat.[11] A very dark coloured lizard with a long rounded tail probably describes the *wandie* (king skink), which can grow to over 50 cm long and lives around rock outcrops that have crevices, rock slab shelters and dense vegetation. Another species described, up to 45 cm in length with short legs, large scales, wide mouth and purple tongue displayed when threatened, was the *youern* to Nyungar and *yurna* to Ngadju (bobtail lizard, *Tiliqua rugosa*); the tail was considered the best part to eat.[11] The fairly common western bluetongue (*Tiliqua occipitalis*) is also described as Aboriginal food. Some sources claim only non-poisonous snakes were eaten by most Aboriginal groups, including the 2.5 m southern carpet python (*wacku* or *bultha*, *Morelia spilota* subsp. *imbricata*); others believe most poisonous snakes were also eaten.[27]

Fish and other seafood are constantly referenced by early European observers as being of great importance to coastal and riverine Aboriginal peoples. The Murray River peoples, such as the Bangarang, describe how important the Murray River cod and other river fish were as a dietary staple.[32] Though lacking the larger river systems of eastern Australia, the smaller rivers, their estuaries and rocky ocean headlands provided abundant fish and other seafood. Aboriginal people fished in inlets, estuaries and rivers for mullet, silver bream and white bait. Fish were driven into nets made of wire grass and, when captured, were placed in bark vessels, taken back to camp and roasted in ash by the women. In the Esperance bioregion, trips were made to Esperance Bay where pink snapper, blue groper and other large fish were plentiful. They were caught by baiting the waters around rocks with broken-up crabs and shellfish to attract them; the feeding fish were then speared off the rocks.[23]

In 2012, traditional Esperance Nyungar owners, with the aid of archaeologist David Guilfoyle and other researchers, rediscovered stone remains at Gabbie Kylie near Esperance that appeared to be a fish trap across a tidal creek and estuary. As has been reported for other regions, the stones would have supported sticks and brush placed across the creek to act as weirs and trap fish as the tide receded. Local Aboriginal people are also reported to have made scoops from shrub branches to rake or scoop fish out of traps such as these.[33] When excess fish were caught, they may have been stored temporarily as Nind observed at Albany, 'they often kill more than is sufficient for present use; in this case, they roast them, and separating the flesh in large flakes from the bones, pack it carefully up in soft bark, in which way it will keep good for several days'.[28]

Insects and other invertebrates were important food resources at different seasons and some, such as various types of beetle larvae (*bardi* grubs) were a highly regarded delicacy. Others included locusts, grasshoppers, termites, ants, beetles and caterpillars. Consumable insects can have a high protein and fat content and may have contributed significantly to the total yearly food intake. Sandra Bukkens' paper on the nutritional value of edible insects throughout the world includes detailed analyses of the nutritional value of many Australian species. Australian witchetty grubs (*bardi* grubs) were found to be not only high in protein (13.2%) but even richer in fat (36.2%) than most species throughout the world.[34] Hassell describes termites and ants, which are even higher in protein and fat, being eaten at the egg and larval stages at Jarramongup.[7]

Plants

The high animal protein diet consumed by Aboriginal peoples was supplemented by a proportion of vegetable food to provide carbohydrates, various vitamins and minerals and roughage. Even when animal protein was abundant, vegetable food was eaten as an accompaniment. In Western Australia, the heathlands are rich in plant foods throughout the year, with many snack foods of small fruits and green herbs such as bower spinach (*Tetraginia implexicoma*) and sea celery (*Apium prostratum*). In December 2016 and January 2017, I harvested the delicious small pink and brown fruits of the *chuck* trees (*Exocarpus sparteus* – native cherry, broom bush) as it was fruiting abundantly across the Esperance sandplain and along the coast to Albany. There were such large amounts of fruit that despite my inexperience I found them easy and fun to strip into a wide basket. I sprinkled them on peanut butter toast, and found them delicious. Hassell describes women harvesting this food by spreading their cloaks under a tree and shaking the fruit off.[7]

Plants high in carbohydrates, vitamins and minerals included pigface (*Carpobrotus virescens*), indigenous bulrush species (*Typha domingensis and T. orientalis*), macrozamia fruits (*Macrozamia dyeri*), grasstrees (*Xanthorrhoea platyphylla*), yam species and orchid tubers.[35,iii] Claude Riche, official botanist of the D'Entrecasteaux expedition, was lost for a period near the site of modern Esperance town and Pink Lake. He described plant foods that Aboriginal people were eating, as similar to 'Hottentot bread' of South Africa – these are yams of the family *Dioscorea*.[36,37] Yam daisy (*Microseris scapigera/lanceolata*) is rare in the Esperance bioregion though still found in refugia around salt lakes, and may have been one of the important yams that Tom Dimer described as disappearing over time.[23] Botanist Beth Gott has concluded that its dramatic decline in eastern Australia was caused by overgrazing pressure mainly by sheep, and that it disappeared within 10 years of European settlement in Victoria and New South Wales.[38] Its destruction by the mid-1840s is linked to the disappearance of the once widespread white-footed rabbit-rat (*tchuteba, Conilurus albipes*). There was starvation among Aboriginal groups including the Wurundjeri, as yam daisy and rabbit-rats disappeared. It was the staple plant diet for both rabbit-rats and people. It seems that Aboriginal people, yams and rabbit-rats were a beautiful example of an interdependent ecoagricultural system, with people especially providing a cultivating disturbance to ensure abundance of the yam. The *tchuteba* was dependent on these managed yam fields and could be viewed as a semi-domesticate providing the staple meat supply for the people, who in return provided them with habitat and food.[39]

---

iii Keighery and McCabe's research supports *T. orientalis* as being indigenous to Western Australia.

Aboriginal people's desire for carbohydrate and the difficulty experienced in easily obtaining it, is reflected in the readiness with which flour and rice became desirable trade goods. George Grey provided the Aboriginal people who worked for him with flour, and they would keenly trade kangaroo meat for it.[12] On Western Australia's southern coast west of Esperance, *youlk* or *youck* (*Platysace deflexa*) was an important staple starchy tuber, as was nardoo (*Marsilea drummondii*), a fern found throughout inland Australia and the eastern Esperance bioregion woodlands in seasonal freshwater bodies. The edible sporocarps of this fern were ground into flour and leached to remove a thiaminase toxin which destroys vitamin B6, then cooked as flat cakes. The deaths of Burke and Wills, who had learnt to eat nardoo cakes provided by Aboriginal people, are thought to have been caused by thiaminase poisoning rather than starvation, as they had failed to understand that the spore flour needed to be leached before consumption.[40] Other important plant foods included ribbon weed (*Triglochin* sp.), which has root tubers and grows seasonally in freshwater; pigface, which grows on the coast and on rocky outcrops – both fruit and leaves were eaten; and bulrush or flag rush (*T. orientalis and T. domingensis*), which produces sago-like starch granules in its roots.[41]

Fire was the major tool for food plant management to clear senescent old growth, stimulate and synchronise flowering and fertilise with ash. This was done for bulrush, orchids, grasslands, grasstrees, macrozamia and many other species. The second way to manage plant foods involved various forms of cultivation using specifically designed tools to till the soil and harvest from special parts of the landscape that grew tuberous plants. The yam daisy throughout southern and eastern Australia, the *Dioscorea* yams on the west coast of Western Australia around Dongara and the *youlk* found on the Western Australian south coast were cultivated in this way. These important root and tuber plants are adapted to needing soil disturbance to persist and proliferate. Geoff Goodall, who is involved in the development of *youlk* for commercialisation, believes that fire use alone will not keep these plants in the landscape as their production and persistence requires yearly disturbance and harvesting to prevent extinction.[42]

Most native fruits eaten throughout Australia were not like modern western fruits, where commercial selection pressure has made the edible part of the fruit sweeter and much larger in proportion to the seed, at the expense of nutrient concentration. Native fruits have much higher levels of roughage, vitamins and minerals, allowing far smaller portions to give health and nutritional benefits.[18] As many native plants contain toxic compounds evolved to combat herbivores with natural pesticides, poisons and unpleasant tastes, safe human consumption relies upon knowledge of how to process them, usually through leaching and fermenting to neutralise toxins. Leached and fermented macrozamia fruits, though not the toxic nuts, were a valuable Western Australian food, though in eastern Australia only the nuts were leached and processed.[43] Given the antiquity of human–cycad interactions, the fruit treatment techniques are likely ancient, as cycads occur in many regions beyond Australia. Methods of detoxification were likely known by the first settlers over 50 000 years ago. Grey observed the detoxification methods carried out by women who collected the fruit in March, and left them to soak in water for several days. The fruits were then buried in rush-lined holes in dry sandy ground to an arm's depth and covered with grasstree leaves. They were left to completely dry before being eaten, either raw or roasted.[12]

It is unknown how detoxification methods were developed for edible plant species – whether through insight and extrapolation, or accidentally with trial and error based on observation. The consumption of clay and soil (geophagy) as a protection against toxins is practised by many animal species as well as humans.[44,iv] It appears to have been well known among the Esperance bioregion peoples, as ethnologist Carl von Brandenstein has described people mixing soil and clay materials with food plants as a storage and transportation method.[45] Grey described Aboriginal people mixing clay with the hot and spicy, red *Haemodorum coccineum* root to make it more palatable.[12] There may have been some level of physiological adaptations to toxins.

Aboriginal people could exploit toxic plants indirectly by eating the herbivorous animals adapted to eat them. For example, common bronze-wing pigeons (*Phaps chalcoptera*) can eat seed of the abundant and widespread *Gastrolobium* species that contain the deadly poison sodium fluroacetate (1080). They were a good food source, as long as the toxin-containing bones and organs were not eaten.[46]

Contrary to general opinion, Aboriginal people throughout Australia had a variety of storage systems for a range of plant foods. Grains were stored in wooden containers and hidden in caves or hollows and covered with bark slabs. Botanist Tim Low describes the discovery in New South Wales of wooden containers storing up to 1000 kg of grass seed.[47] Aboriginal plant foods specialist and botanist Philip Clarke reports descriptions of nardoo sporocarps, various grass and wattle seeds and a range of fruits being processed and stored throughout Australia. For instance, nardoo sporocarps were stored in woven bags and stored in caves or dry sand; quandong (*wolgol* in Western Australia) fruit was dried and stored; bush tomato was threaded on clean sticks, dried and stored; other fruits when abundant could be mashed, dried into cakes and stored. Dried fruits were also used as trade items.[48]

Foods could be processed with edible preservative material such as gum from acacias and eucalypts, and stored for long periods. At Jarramongup, Hassell describes how soft acacia gum (*meen*) was collected by the women and formed into large balls that hardened. When wanted for eating, lumps were knocked off the ball and heated to soften and break off pieces. She also describes *quonert*, which was mixture of black wattle and jam tree seed ground into an oily meal then mixed with yate sap (like molasses in taste and texture) and baked in ashes. Large amounts of the seed were collected communally, winnowed then stored in the women's collecting bags to be used communally over time.[7]

Storing food as living animal and plant environmental biomass was arguably the most efficient overall solution that Aboriginal people developed to overcome technological difficulties in food storage. This was accomplished through their many ecologically based cultural systems that promoted availability and abundance well into the future. In contrast to the Aboriginal system of future-proofing, subsistence farming can be hard and unrelenting work – farmers have turned harvesting from nature into a battle with nature as they strive to

---

iv Geophagy is common in animal species, so is likely to have been used by humans in antiquity.

maintain highly simplified ecosystems in early successional stages, often in places where the food plants or animals do not naturally grow.[49] Hassell's descriptions of foraging with the women of the Wheelman group at Jarramongup in the 1880s portray an enjoyable social occasion for those involved as they harvested predictably from their managed ecosystems.[7]

Water

Water distribution and supply were vital parts of Aboriginal SES, for the survival of people and their animals and plants depended not only on its natural availability in the landscape, but also upon whether people could increase its availability. Evidence of long occupation can be found at almost any rockhole, soak, freshwater lake or river system. Scatterings of stone flakes are often found in sandy areas around these water sources where Aboriginal people camped. At Coronet Hill there is a flint-laden ridge above a 200 ha lake – a favoured place to sit and make flint tools, maybe while planning hunts of water birds from the lake below, or of the bandicoots, brushtail and tammar wallabies living in the surrounding paperbark swamps and dense thickets.

If water was not available in an area, it meant that the resources there could only be exploited for short periods. Water scarcity limited the animal numbers that an area could sustain, and the length of time people could stay. Interventions that ensured water quantity and duration for humans, wildlife and edible water plants were devised. In granite country, the seasonally transient shallow water bodies that collect on and around the base of the granite hills and outcrops, some held in deep rock fissures, were improved.[50] Channels and fissures were blocked with clay and stones to form small deep dams which held rainwater for long periods. Water was also conserved in *gnammas*, human-deepened depressions and holes on granite outcrops. These initially form as rock depressions, collecting organic matter that rots and produces granite-eroding carbonic acid. The rotting rock material was scratched out by animals as well as being cleaned out periodically by Aboriginal people to enlarge and deepen the *gnamma*. Fire was used to further enlarge and deepen the *gnamma* as a long-term ongoing process. I have found *gnammas* ranging from a few centimetres to many metres deep, with some containing hundreds of litres of water.[50] In the past, *gnammas* were vital sources of water for the Ngadju in the mallee and woodlands of the Esperance bioregion. Aboriginal people placed sticks and branches in *gnammas* to allow mammals and birds to drink and climb out without drowning. *Gnammas* were of lesser importance in coastal areas and sandplain as numerous streams, freshwater swamps and soaks supplied permanent water.

The granite domes in the sandplain rarely feature deepened *gnammas* but often have waterholes and springs around the base at shallow depth in deep sands likely dug out by people. Providing permanent water access to animals allowed greater numbers of them to live in an area where food was not limiting, compared to places without accessible water. Grey kangaroos have been observed to scrape depressions in damp areas to uncover water on Coronet Hill, during summer. This habit has been described by ecologist Terry Dawson as important in providing water to kangaroos and other species.[51,v]

---

v   Dawson describes watching a euro dig a waterhole in a dry creek bed, watched by a group of goats, a fox and several rabbits.

Ecoagriculture for a Sustainable Food Future

Fig. 5.1: Pair of gnammas in a rock outcrop west of Balladonia Road, Shire of Dundas, Western Australia.

In the woodlands without natural water sources, Ngadju people developed a system of 'water trees', called *pillirri* around Balladonia and *kumbal* around Fraser Range. Making a water tree shows people's long-term commitment to living in their Country and helping future generations. It was a very long-term process but once completed the *pillirri/kumbal* tree would last for many hundreds of years or the life of the tree. The process involves carefully training a multi-stemmed tree by packing sticks and clay into the area where the stems meet; over time, the tree grows to form a large deep bowl. During rain events, the water runs down the stems into the bowl, which can then be covered with bark to prevent evaporation and keep animals out of it. A water tree was usually a salmon gum (*marrlinja*, *Eucalyptus salmonophloia*) or a black morrel (*pilerli*, *Eucalyptus melanoxylon*). Water trees are particularly found along travel routes and are still in existence today. Other types of water trees, such as kurrajongs (*Brachychiton populneus*) and Christmas tree (*kunapiti*, *Grevillea nematophylla*) stored large amounts of water in their thick watery roots. Several eucalypts such as the giant mallee (*E. oleosa*) and black morrel (*E. melanoxylon*) also have roots that store significant amounts of water.[52] Tom Dimer explains how he had been taught to find water, 'There's water in the bulloak trees ... the desert bull oak (*Allocasuarina* species). And there's water in the kurrajong trees and they got a bowl of water ... You can get any amount of water'.[23]

Edward John Eyre's observations throughout Australia provide further evidence of Aboriginal peoples' intimate forward planning and necessary knowledge of water sources. He noted that Aboriginal people were likely to think ahead and procure water from dew and roots in dry areas well before they needed it. He compared this to the short-term outlook of most Europeans, who waited until they were in dire straits before desperately looking for water.[53]

Embedding food and culture in nature and Country

When the original human settlers came to Sahul in the deep past, the seemingly endless abundance of readily exploitable food animals and plants would have made restraint seem unnecessary. When animal populations collapsed in distribution and abundance, the human population would have experienced tremendous social upheaval and hunger, with growing food crises and regional climatic and environmental changes. These difficult times may be reflected in the battle scenes so often depicted in the ancient Gwion Gwion paintings of the Kimberley.[54] What is so remarkable is that after this collapse, Aboriginal people throughout Australia went on to develop sustainable and resilient SES that, despite missing the suites of wonderful prehuman animals, created a new version of Australian ecosystems. The material culture of the Nyungar, Ngadju and other Australian Aboriginal peoples has been described as simple, but a more accurate view is that it was exquisitely designed and crafted to meet the needs of its users most efficiently without overextraction from interdependent ecosystems.

The fundamental necessity of long-term ecological thinking and its practical implementation for food production and resource use is largely ignored in 'civilised' societies that are based on short-term economic and social cycles.[55] Aboriginal livelihoods, individually and as groups, depended directly on everyday intellectual thinking incorporating ecology – the seasonal nuances, the lifecycles of animals and plants and, crucially, how people needed to live in order to ensure and enhance these interrelationships for their future. Culture and religion throughout Australia were based on lessons, warnings and avoiding past troubles in the Dreaming. They acted to ensure a continuity of ecoagricultural knowledge and relationships with Country.

Endnotes

1. Teicholz N (2014) *The Big Fat Surprise*. Scribe, Melbourne.
2. Goldacre B (2008) *Bad Science*. Harper Collins, London.
3. Malhotra A (2013) Saturated fat is not the major issue. *British Medical Journal* **347**, f6340. doi:10.1136/bmj.f6340
4. Ottoboni A, Ottoboni F (2004) The food guide pyramid: will the defects be corrected? *Journal of American Physicians and Surgeons* **9**(4), 109–113.
5. Martin DL, Goodwin AH (2002) Health conditions before Columbus: paleopathology of native North Americans. *Western Journal of Medicine* **176**(1), 65–68. doi:10.1136/ewjm.176.1.65
6. Mann N (2007) Meat in the human diet: an anthropological perspective. *Nutrition & Dietetics: Journal of the Dietitians Association of Australia* **64**(suppl.4), S102–S107. doi:10.1111/j.1747-0080.2007.00194.x
7. Hassell E (1936) *My Dusky Friends*. C.W. Hassell, Perth. Reprinted 1975.
8. Yen AL (n.d.) *Edible Insects and Other Invertebrates in Australia: Future Prospects*. Biosciences Research Division, Department of Primary Industry, Melbourne.
9. Cunningham I (2005) *The Land of Flowers: An Australian Environment on the Brink*. Otford Press, Sydney.
10. Graham S (2016) Aboriginal Elder of the Ngadju Nation. Interviewed by Nicole Chalmer, Coronet Hill.

11. Dimer D (2016) Member of the Ngadju Nation. Interviewed by Nicole Chalmer, 5 November.
12. Grey G (1841) *Journals of Two Expeditions of Discovery in North-West and Western Australia, during the Years 1837, 1838 and 1839, under the Authority of Her Majesty's Government, Describing many Newly Discovered Important and Fertile Districts, with Observations on the Moral and Physical Condition of the Aboriginal Inhabitants*. Vol. 2. T. and W. Boone, London.
13. O'Dea K (1991) Traditional diet and food preferences of Australian Aboriginal hunter-gatherers. *Philosophical Transactions of the Royal Society of London. Series B, Biological Sciences* **334**(1270), 233–241. doi:10.1098/rstb.1991.0112
14. Smith RM, Smith PA (2003) An assessment of the composition and nutrient content of an Australian Aboriginal hunter-gatherer diet. *Australian Aboriginal Studies* **2**, 39–52.
15. Brand-Miller JC, Colagiuri S (1994) The carnivore connection: dietary carbohydrate in the evolution of NIDDM. *Diabetologia* **37**(12), 1280–1286. doi:10.1007/BF00399803
16. Meagher S (1975) *The Food Resources of the Aborigines of the South-west of Western Australia*. WA Museum, Perth.
17. Brand-Miller C, Holt SHA (1998) Australian Aboriginal plant foods: a consideration of their nutritional composition and health implications. *Nutrition Research Reviews* **11**, 5–23. doi:10.1079/NRR19980003
18. Newton J (2016) *The Oldest Foods on Earth: A History of Australian Native Foods – with Recipes*. NewSouth Publishing, Sydney.
19. Prober SM, O'Connor MH, Walsh FJ (2011) Australian Aboriginal peoples' seasonal knowledge: a potential basis for shared understanding in environmental management. *Ecology and Society* **16**(2), 12. doi:10.5751/ES-04023-160212
20. McDonald H (2003) The fats of life. *Australian Aboriginal Studies* **2**, 53–61.
21. Brightman R (1996) The sexual division of foraging labour: biology, taboo and gender politics. *Comparative Studies in Society and History* **38**(4), 687–729. doi:10.1017/S0010417500020508
22. Hallam SJ (1975) *Fire and Hearth: A Study of Aboriginal Usage and European Usurpation in South-western Australia*. Advocate Press, Melbourne.
23. Dimer T (1990) *Outback Station Life; Aboriginal Customs and Beliefs; Bush Skills and Survival; Vermin Control for Agricultural Protection Board*. OH 2339 – Interviewer, Helen Crompton (transcription). Oral History Unit, J.S. Battye Library, Perth.
24. Smith M (1993) Recherché a l'Esperance: a prehistory of the Esperance region of south-western Australia. PhD thesis, University of Western Australia, Perth.
25. Hannett E (2007) Retired farmer. Interviewed by Nicole Chalmer, Esperance, 19 January.
26. Kerle A (2001) *Possums: The Brushtails, Ringtails and Greater Glider*. University of NSW Press, Sydney.
27. Dimer K (1989) *Elsewhere Fine*. South West Printing and Publishing, Bunbury.
28. Nind IS (1831) Description of the natives of King George's Sound (Swan River colony) and adjoining country. *Journal of the Royal Geographical Society of London* **1**, 21–51. doi:10.2307/1797657
29. Palomares F, Gaona P, Ferreras P, Delibes M (1995) Positive effects on game species of top predators by controlling smaller predator populations: an example with lynx, mongooses, and rabbits. *Conservation Biology* **9**(2), 295–305. doi:10.1046/j.1523-1739.1995.9020295.x
30. Ritchie EG, Johnson CN (2009) Predator interactions, mesopredator release and biodiversity conservation. *Ecology Letters* **12**(9), 982–998. doi:10.1111/j.1461-0248.2009.01347.x
31. Daisy B (1936) *My Natives and I – incorporating The Passing of the Aborigines: A Lifetime spent among the Natives of Australia*. (Ed. PJ Bridge). Hesperian Press, Perth.
32. Bickford A (1982) *Our People*. Methuen, Sydney.
33. Guilfoyle D (2012) *Stone Fish Trap Rediscovered*. Gabbie Kylie Foundation, National Trust of Australia and Applied Archaeology, Esperance.
34. Bukkens SGF (1997) The nutritional value of edible insects. *Ecology of Food and Nutrition* **36**(2–4), 294–296.
35. Keighery G, McCabe S (2015) Status of *Typha orientalis* in Western Australia. *Western Australian Naturalist (Perth)* **30**(1), 30–35.
36. De Rossel M (Ed.) (1808) *The Voyage of D'Entrecasteaux*. Imperial Press, London.
37. Hoggart M (2014) Farmer and botanist. Interviewed by Nicole Chalmer, Condingup, 10 November.
38. Gott B (1982) Ecology of root use by the Aborigines of southern Australia. *Archaeology in Oceania* **17**(1), 59–60.

39. Mansergh I, Cheal D (2019) Connecting the dots of simultaneous extinctions: 'Tchuteba', yam daisies and Aboriginal cultivation. *Victorian Naturalist* **136**(2), 78–84.
40. Stipanuk MH, Caudill MA (2013) *Biochemical, Physiological, and Molecular Aspects of Human Nutrition.* 3rd edn. Elsevier, New York.
41. Woodall GS, Moule ML, Eckersley P, Boxshall B, Puglisi B (2010) *New Root Vegetables for the Native Food Industry: Promising Selections from South-Western Australia's Tuberous Flora.* Publication 09/161. Rural Industries Research and Development Council, Wagga Wagga.
42. Goodall G (2014) Consultant botanist. Interviewed by Nicole Chalmer, Albany, 17 July.
43. Macintyre K, Dobson B (2018) *The Ancient Practice of Macrozamia Pit Processing in South-western Australia.* <https://anthropologyfromtheshed.com/project/the-ancient-practice-of-macrozamia-pit-processing-in-southwestern-australia/>
44. Rowland MJ (2002) Geophagy: an assessment of the implications for development of Australian Indigenous plant processing technologies. *Australian Aboriginal Studies* **1**, 55–59.
45. von Brandenstein CG (1977) Aboriginal ecological order in the south-west of Western Australia: meaning and examples. *Oceania* **47**(3) 169–252.
46. Peacock DE (2003) The search for a novel toxicant in *Gastrolobium* (Fabaceae: Mirbelieae) seed historically associated with toxic native fauna. PhD thesis, University of Adelaide, Adelaide.
47. Low T (1991) *Wild Food Plants of Australia.* Harper Collins, Sydney.
48. Clarke PA (2011) *Aboriginal People and Their Plants.* Rosenberg Publishing, Sydney.
49. Gliessman SR (2007) *Agroecology: The Ecology of Sustainable Food Systems.* 2nd edn. CRC Press, Boca Raton.
50. Bindon PR (1997) Aboriginal people and granite domes. *Journal of the Royal Society of Western Australia* **80**, 173–174.
51. Dawson T (1997) Desert kangaroos. In *Windows on Meteorology: Australian Perspective.* (Ed. EK Webb). CSIRO Publishing, Melbourne.
52. Prober SM, Yuen E, O'Connor M, Schultz L (2013) *Ngadju kala: Ngadju Fire Knowledge and Contemporary Fire Management in the Great Western Woodlands.* CSIRO Ecosystem Sciences, Perth.
53. Eyre EJ (1845) *An Account of the Manners and Customs of the Aborigines and the State of their Relations with Europeans.* T. and W. Boone, London.
54. Wilson I (2006) *Lost World of the Kimberley.* Allen and Unwin, Sydney.
55. Whiteman G, Cooper WH (2000) Ecological embeddedness. *Academy of Management Journal* **43**(6), 1265–1282.

6
Overrun by sheep: the pastoral template for colonisation

> The game of the wilds that the European does not destroy for his amusement are driven away by his flocks and herds … The waters are occupied and enclosed, and access to them is frequently forbidden.[1]

Edward John Eyre was one of the few European explorers who recognised the prior land occupation and ownership by Aboriginal peoples, and expressed deep concern about land alienation, their displacement and the impact of poorly managed European herbivores. He lamented that the country most suited for cultivation and grazing was also the most valued by Aboriginal peoples for food production.[1]

The Anglo/European colonisation of much of Australia commenced with linked animal–human systems that enabled the invasion and occupation of Aboriginal lands. Environmental historian Alfred Crosby's concept of ecological imperialism describes how large numbers of domestic herbivores helped invading colonists throughout the world take over new landscapes from indigenous peoples.[2] In her book *A Plague of Sheep*, Elinor Melville traces this process in Mexico in the 16th century when the Spanish claimed the land from indigenous peoples, helped by sheep as biological co-invaders.[3] Similar mechanisms were used throughout much of the Australian frontal wave of invasion, as large herds of sheep and cattle helped colonists expand their range.

Early pastoralists in the Esperance bioregion used their sheep flocks to aid invasion of the sandplain-mallee lands of the Nyungar and sandplain-mallee-woodlands of the Ngadju Aboriginal peoples. Because the Europeans initially used locally adaptive behaviours, including mobile eco-shepherding learnt from Aboriginal people and modelled on the ancient European transhumance system, their early environmental impacts were not as immediately devastating as in most of Australia. Yet within three decades, the cultural drive for economic growth and a lack of critical understanding about impacts of excess sheep stocking rates facilitated an ungulate eruption that contributed to the collapse of animal–plant ecosystems and Aboriginal social ecological systems (SES).

Drivers of invasion and colonisation

The first shipboard explorers who sailed, mapped and named the coast and islands of the Recherche Archipelago along Western Australia's south coast were the French. The long period of wars against England had resulted in France losing most of its colonial territories, so there was a push to explore the southern lands with a view to expand scientific knowledge, set up trade and perhaps colonise the western coast of Australia.[4] From an ecological viewpoint, European invasion was also linked to the environmentally unsustainable SES of Britain and other European societies. Based on unrestrained economic and population growth, they were outgrowing their ecological limits – by 1788 Great Britain was no longer self-sufficient in food production.[5] Britons' diet was dominated by cheap nutrient-poor carbohydrates, such as wheat and potatoes, with meat a rare luxury. As John Burnett outlines, industrialisation and the burgeoning population meant that food self-sufficiency was impossible for most. There were periods of starvation caused by unemployment and poverty, linked to population growth, economic cycles and the practice of exporting food for higher monetary returns rather than keeping it to feed the country's poor.[6]

By the time the British claimed the lands of the Swan River Nyungar people in 1829, the practice of taking communal lands and concentrating them into individual ownership was historically well established. In Britain, the procedures of Enclosure during the 14th to 17th centuries took land from the many and gave it to a few wealthy nobility.[7] Enclosure meant that individual families were no longer able to live autonomously and feed themselves from their own plots and common lands – they were forced into slave-like industrial employment, starvation or migration. Before Enclosure, common lands were managed skilfully and (in a striking parallel to Aboriginal SES) communally, with group consensus decisions made by the long-term occupants.

Earlier invasions of indigenous lands in other parts of the world had set a precedent for disenfranchisement. As historian James Blaut proposes, there was a colonial 'myth of emptiness' about indigenous cultural institutions and how people lived in and managed the invaded lands, unrecognised by British legal systems of land ownership. Blaut considers this myth a prerequisite for the immorality of colonial invasion and settlement, for such lands were not overpopulated as much of Europe had become.[8] Places in Australia such as the Esperance bioregion were regarded as nearly empty of people; therefore, settlement by Europeans was not generally assumed to displace Aboriginal people. Widely perceived as wandering nomads, with no sovereignty or claim to territory, Aboriginal peoples were incorrectly judged to have no concept of private property ownership or productive agricultural practices such as plant cultivation and domesticated animals.

Though there are numerous examples of individuals recognising that Aboriginal peoples had forms of inherited land and hunting ownership, along with sophisticated land management practices, it was expedient for colonial governments and entrepreneurial developers not to acknowledge this. Without markers of 'civilisation' such as buildings and ruins, no obvious evidence of soil cultivation, and with no land titles recognised by British law, the land was deemed *terra nullius*.[9] Under British law the Crown became the landowner, and the colonial administration could grant, sell or lease it to European settlers.[10] This outlook had already been exhibited in North America where, as Dennis Blanton explains, early British colonisation was further validated by a belief in divine right and even an 'obligation' of English people to bring the wilds of the New World into production by imposing an English and European model of agriculture and civilisation upon the invaded country and people.[11]

British society further justified the seizure of indigenous lands with stage theory.[12] Stage theory asserted the superiority of one society over another in temporal terms with an underlying belief in racial hierarchy and social progress. It posited that humanity evolved through a natural sequence of progressive developments, from 'primitive' hunter-gatherers, to herders, to the agrarian cultures of Europe and thus civilisation. Societies deemed 'primitive' were thought to have remained in the hunting stage because they had not been subjected to pressures such as overpopulation and consequent land shortages, or because they had no suitable native animals or plants for domestication. It was believed they were inevitably doomed in the face of a 'superior' culture and race. From an ecological viewpoint this interpretation can be turned on its head, as European and other diasporas can be interpreted as largely driven by unsustainable cultural practices such as overpopulation, environmental overexploitation, economic growth reliant upon consumerism, technological advancement, and population growth to consume stuff in the boom–bust cycles that continue today. This drove an endless need to expand territory. In contrast, many indigenous lands were resource-rich as they were managed relatively sustainably by the inhabitants, who were far ahead in their cultural understandings that linked restraint to environmental sustainability. The resources they had fostered for generations were seized by the invaders and rapidly exploited and depleted.

The common European/Anglo attitude to native animals and plants when invading indigenous lands was extractivist (exploitable for money and amusement) and shaped by the belief that their own domesticates and wildlife were superior for food, civilisation, making money and as a familiar landscape presence.[13] The theory of evolution by natural selection developed by Charles Darwin and Alfred R. Wallace in the mid-19th century was interpreted as clear evidence to support those beliefs.[14] The European attitude towards Aboriginal people and their culture carried over to Australian marsupials – they were considered 'primitive' predecessors of modern placental mammals, doomed as the more 'advanced' domestic and introduced mammals were established. This view was advocated by zoologist Ernst Haeckel, who drew phylogenetic 'trees of life' that displayed his Darwinian theories that 'man' was the evolutionary pinnacle and marsupials were primitive ancestors.[15] This flawed scientific view still had some influence in the mid to late 20th century. Zoology lectures I attended for my science degree in the late 1970s–1980s continued to describe marsupials as primitive mammals surviving only because of Australia's isolation and the absence of placental mammal competitors.

European settlement of eastern Australia soon resulted in marsupials being considered as serious competitors to livestock. They were declared vermin, and policies for their extermination, mostly with strychnine baits spread throughout the landscape, were enacted by the 1880s in all states in eastern Australia.[16] In 1887 the Queensland Parliament passed an *Act to Facilitate and Encourage the Destruction of Marsupial Animals*. By 1930, under various versions of that Act, over 27 million animals had been killed (mostly kangaroos, wallaroos, wallabies, pademelons, kangaroo rats, bandicoots, dingoes and foxes) in Queensland alone, confirmed by scalp bounty records.[17] This number is vastly underestimated as many were killed but not recovered. It also does not include those killed for fur or skins. Similar Acts were passed in New South Wales and Victoria, where further millions of animals were killed. These killings, and the destruction of habitat, would have had profound effects on the extinctions of marsupial species.

In Western Australia extermination was not so fervent, though strychnine was used very early in European colonisation to exterminate dingoes, emus and eagles that threatened sheep. It was recognised that marsupial herbivores were essential food animals for Aboriginal peoples, and various Game Acts were enacted in the 1890s to protect a selection of species.[18]

Early economic expansion in Western Australia and other colonies depended on extraction of natural resources and direct appropriation of the products of nature, such as fur skins from seals and possums, and meat and hides from kangaroos and wallabies, to turn into monetary profits.[19] The unique abilities of native animals such as kangaroos and emus, to survive and prosper on native grasses, shrublands and woodlands with little water or need for coast to inland stock movement (transhumance) in order to get adequate minerals needed by domesticates, was never seen as a sustainable opportunity, nor as examples from which to learn and embed new types of grazing systems. These low-cost sustainable opportunities are still largely unrecognised, as the dominant Australian food culture is based on the deeply cultural belief that beef and sheep meat are really the only 'acceptable' red meats.

Aboriginal landscapes

The Esperance bioregion encompassed Aboriginal SES developed over 40 000–60 000 years of Aboriginal land management and ecosystem co-evolution.[20] This is the basis for how people lived in the Esperance bioregion and throughout Australia in both pre-and post-colonial times.

In 2013, Susan Prober and partners investigated the many ways Ngadju people ecologically managed their lands for food production. A complex and elegant system of fire use created and maintained mosaics of managed habitats for important animals and plants in ways that are still not fully understood. Periodic burning of habitats was vital for pioneer type plants such as orchids, typha and seed-producing grasses, and for food animals such as kangaroos, wallabies, smaller marsupials and birds such as emus, bustard (*Ardeotis australis*) and other grassland species that need nutritious new growth. Larger vegetation and trees were actively protected from burning by only firing under ideal conditions and by using branches to beat out flames that got too close, as the trees were a habitat 'city' for a variety of animals.[21]

As the sustainability of Aboriginal SES could be easily compromised by resource overexploitation, the groups used various restraining approaches to prevent this. As previously discussed, they managed people (themselves) as sustainable populations with a wide range of cultural strategies, and managed nature with a variety of plant and animal protection methodologies.[22] As Johan Colding and Carl Folke confirm, these actions are common worldwide among indigenous peoples, and have prevented the many possible fluctuations in animal and plant populations that could result from an excessive human population overexploiting an ecosystem.[23] Interactions between people and nature seemed designed to promote food abundance and biodiversity as the normal resilient and sustainable state.

During his 1848–49 expedition to the Esperance bioregion, Western Australia's State Surveyor John Septimus Roe observed:

> … the coastal land appeared closely settled, with Aborigines, numerous kangaroos and emus occurring together in this same country. There was much smoke from native fires and the occurrence of native huts was common.[24]

He described large patches of grasslands distributed along the coast and further large patches in the mallee woodlands. The original native grasslands included tall warm-season perennial grasses such as kangaroo grass (*Themeda triandra*), wallaby grass (*Austrodanthonia caespitosa*) and spear grasses (*Stipa eromophila*, *Austrostipa nitida*). There were bi-annuals such as windmill grass (*Chloris truncata*), edible herbs, including pigface (*Carpobrotus* sp.) and shrubs such as bluebush (*Maireana* sp.) and saltbush (*Atriplex* sp.).[25] Aboriginal people as ecofarmers had developed these grasslands for their native herbivores to graze; unfortunately, they attracted Anglo pastoralists with their herds of sheep, cattle and horses.

Ungulate eruptions

When a species, whether it be human, domesticated herbivore, wild animal or plant, invades or spreads into a new range with unlimited food and lack of predators, a runaway population growth can erupt. This means the population explodes to a peak, crashes and then has a carrying capacity lower than its peak because of overexploited resources and irreparably damaged ecosystems. In 1970, population ecologist Graeme Caughly referred to this paradigm when related to escalating numbers of ungulate herbivores, such as deer, sheep and cattle, as an 'ungulate eruption'. In natural systems, invasion into new unexploited territories frequently results in runaway population growth.[26] During the European diaspora, domestic herbivores such as cattle and sheep were encouraged in this process by their human co-invaders. In his discussions about ecological imperialism, environmental historian Alfred Crosby describes how invading colonists deliberately or inadvertently harnessed this animal behaviour feature and assisted the eruptions and spreading invasion of their domestic grazing animals, to physically, psychologically and culturally claim lands from indigenous peoples.[2]

Elinor Melville states that this human–animal co-invasion process aided the Spanish when they claimed Mexico from the indigenous Mexicans in the 16th century. With the help of vast spreading sheep herds that consumed indigenous crops, grass fields and ecosystems and ultimately the human culture, people were effectively disenfranchised. However, the ruinous overgrazing event also affected the Spanish sheep farmers, as the severely degraded landscapes would no longer carry the previous numbers of sheep – and still cannot.[3] This pattern was mirrored in much of Australia as environmental destruction caused by eruptions of sheep and cattle in rangelands, especially around permanent water sources, dramatically and permanently reduced the land's carrying capacities to this day.[27] The impact was described very early by Sir Thomas Mitchell who visited the New South Wales Bogan River valley first in 1835, then in 1845. He was dismayed at how, in only 10 years, excessive livestock had trampled the once beautiful springs, destroyed the surrounding wetland vegetation and eaten the valley bare so that 'hardly a blade of the once verdant grasses were to be seen'.[28] This could also describe much of the Western Australian rangelands. A station I visited in the Murchison River region had once carried over 90 000 sheep, bringing vast monetary returns from wool. It has an enormous stone shearing shed and a crumbling, vast luxurious brick and stone homestead with once elegant gardens, surrounded by unoccupied quarters and cottages equivalent to a small town. The property today can only carry a few thousand sheep and several hundred cattle in a severely degraded drought-prone landscape, dominated by inedible shrubs and brief annual pastures.

In the Esperance bioregion and across much of Australia, sheep were the main domestic herbivore used as an ecological colonising agent. Their human-facilitated impact was initially an invasion, as a consumer front of sheep moved throughout the landscape, helping the early

Anglo pastoralists to appropriate Country and replace Aboriginal SES with their own. The early, relatively sustainable numbers and management practices in the Esperance bioregion were followed by an exponential sheep eruption. Rica Ericson describes the Dempster family build-up of sheep numbers at Esperance from the initial 580 in 1865, to 13 750 by 1875 and 35 000 by the 1890s.[29] Other pastoralists also increased their sheep numbers and as many as 60 000–65 000 sheep grazed the native pastures of the Esperance bioregion.

How sheep can change landscapes

Sheep are pro-active animal agents creating and mapping landscapes for themselves with tracks that reflect their daily lives. As Michele Dominy concludes, in New Zealand they 'mediate the colonist's relationships to the [new] land'. They also made the new landscapes navigable by developing pathways linking food and water resources and determining the tracks followed during mustering.[30] In the Esperance bioregion, these tracks likely included more ancient and pre-existing tracks made by Aboriginal people, kangaroos and perhaps extinct megafauna.

Sheep have other behaviours that make them ideal agents of colonisation. They are social animals, especially Merinos which have strong mobbing instincts, and can nearly double in number after a year on good forage when protected from predators. Huge flocks actively invade new grazing areas looking for food, if allowed. At this point, it is worthwhile to discuss the myth of hard hooves so often blamed for ruining Australia's 'fragile untrampled soils'. This concept disregards the soil impacts of the numerous medium to large extinct Australian megafauna. Like most social herbivores (including kangaroos, wild horses and deep past grazers such as diprotodon and vombatid species), sheep tend to form discrete tracks linking food and water resources – they do not randomly stamp all over the landscape. But as human-mediated numbers increase excessively so does grazing, trampling and soil degradation as happens with any overstocked animal, whether ungulate, macropod or human. Rangeland ecologist P.B. Mitchell asserts that the 'common sense' myth about hard hooves being the main reason for soil compaction and land degradation in semi-arid Australia is supported by little empirical evidence, but as an uncritical generalisation, it has become widespread in scientific and popular literature. Consequently, not enough attention is given to more important factors in land degradation, which include excessive stocking rate, impacts of overgrazing, preferential grazing and animal behaviour.[31]

Sheep vigorously graze all palatable herbage within reach, biting off forage with their sharp incisor teeth. Unlike cattle, which use their tongue to wrap around and tear off herbage, sheep can graze to ground level and even below, by digging for roots and seeds with their hooves. They can appear to be healthy even when groundcover is largely gone and soil degradation is accelerating. With planned pasture management using rotational grazing, whereby the sheep flocks or cattle herds are moved to new pastures before overgrazing occurs, palatable perennial grasses and shrubs which are adapted to being grazed and need it for healthy growth, can thrive by having time to recover and regenerate before the next grazing period. The most common form of mismanagement is set stocking, in which too many livestock are left permanently on pastures to choose what, when and where they eat. This works with small numbers but damage accelerates as populations grow. Preferential grazing can annihilate most palatable plants, especially perennial grasses and herbs, thus preventing seedling recruitment and baring ground. These activities lead to soil degradation, the reduction and disappearance of perennial grasslands, dominance of annual grasses and invasion by woody shrubs, unpalatable forbs and sedges. Annual grasses are only present for a short period each year after sufficient rain, and not at all if there are extended dry seasons and the ground becomes bare.

In 1842 John Drummond described how sheep introduced seeds of European and Mediterranean annual grasses and weed species to Western Australia in their dung and wool; these then became naturalised.[32] They included several cool-season annual grasses (*Bromus*, *Hordeum* and *Vulpia* spp.), forbs such as capeweed (*Arctotheca calendula*) and annual legumes (*Trifolium* sp., *Medicago* sp.). These plants, already co-adapted to withstand sheep grazing, can take over bared ground areas and outcompete native perennial grasses, chenopod shrubs and herbs. There are long-term negative impacts on the microclimate, hydrology and soil biology of the region as landscape water retention and availability are reduced, habitats are destroyed, organic matter declines and droughts start occurring again and again.

These deep ecological impacts can lead to desertification as humidifying microclimates deteriorate, for evaporation increases from bared ground as the landscape and ecosystems become annual based. This means that there are long periods when there are no living roots to hold soil together against wind or rain deluges or to provide habitat for beneficial soil organisms. Research in arid to semi-arid regions of the world shows that native vegetation grows in landscape patterns designed to trap rainfall at the habitat level. Without these patterns, a cycle of increasing aridity and drought commences as bioprecipitation decreases and rainfall effectiveness is lost.[33,34] Ground-dwelling animals such as arthropods, small to medium mammals, ground-nesting and seed-eating birds and reptiles are particularly vulnerable as excessive grazing intensity destroys their sheltering habitats, making them susceptible to overheating, starvation and predation.[35] For instance, within 10 years of their introduction into south-eastern Victoria, erupting sheep populations had so wholly dug up and consumed the yam daisy (*murnong*) that Aboriginal people were starving due to the loss of this staple food, and the co-dependent white-footed rabbit-rat became extinct.[36,i] In contrast, another form of mismanagement is too light or no grazing – perennial grasses need intermittent defoliation by grazing to promote regrowth and seed production, prevent senescence and allow fire mitigation.

Mobility, rotation and strong awareness of the plant species necessary in healthy grasslands are the best ways to maintain rangeland productivity.[30] Management practices that include conservative stocking levels and timely rotational grazing, can limit or prevent environmental damage. Ultimately, the actions of people will decide the long-term impacts of grazing by sheep and other domestic herbivores. However, due to the past history of degradation, the need for exceptional management to prevent desertification and their vulnerability to an essential keystone predator, the dingo, it is questionable whether sheep should be allowed in the rangelands at all.

---

i   *Microseris lanceolata* is the new classification name of Gotts *M. scapigera*.

The pastoralists

Pastoral nomadism is based on human and animal mobility. Agroecologists Alireza Koocheki and Stephen Gleissman consider this SES to be a comparatively sustainable way to manage resources and land in arid areas. They conclude that constantly moving grazing animals to new native pastures before overgrazing occurs can be a functional agroecological response to harsh environmental conditions.[37] The British form of pastoralism was usually not adaptive in this way as settlers had little experience with mobile pastoralism as a lifeway. They associated civilisation with sedentary living, considering it superior to the nomadic lifestyle led by gypsies, herders and Aboriginal people.[38] The invaders had an efficiently destructive domestic herbivore colonising template with replicated impacts throughout Australia, as they took over and rapidly overgrazed the existing grazing lands of Aboriginal ecoagriculture. There was also a belief, held especially in eastern Australia, that the introduction of European trees, other plants and livestock, with the action of their hooves, greatly improved not only the productivity but the appearance of Australia. Queensland explorer and writer Ernest Favenc was a firm believer as he described in 1888 how the touch of colonisation appeared 'to electrify the soil':

> Imported animals, trees, and plants lived and flourished among the dingy forests, which barely yielded food enough ... [until] the sure, if gradual change wrought by stocking. Under the ceaseless tread of myriad hoofs, the loose, open soil was to become firm and hard, whilst fresh growths of herb and grass followed the footsteps of the invading herds.[39]

Throughout western, southern and eastern Australia sheep formed the main pastoral base, and the quickest form of income was wool. This gave colonising enterprises a quick cash-paying export commodity and so allowed further expansion.

As access to new agricultural lands around the original Swan River colony declined, the colonial government of Western Australia developed a policy to increase colonial income, consolidate land ownership away from Aboriginal people and expand European settlement by encouraging the pastoral industry to extend into the far reaches of the colony. In the late 1850s it offered generous terms, with leases of 20 000–100 000 acres granted to European settlers, the first four years rent-free. Under these terms, the Dempster brothers from Toodyay took up leases on the south-east coast. They were the first of a wave of Anglo people who invaded and settled in the Esperance bioregion.[29]

The Dempster home farm was based at Northam and Toodyay about 100 km east of Perth. With an erupting family of eight children, five of them sons, the family actively engaged in exploration to acquire new land. They first explored eastward from Northam in 1861 but were unimpressed by the region, though they came to appreciate the importance of granite outcrops as a source of water and feed.[40] Perhaps due to glowing descriptions by colonial Surveyor General J.S. Roe in 1848, an expedition was organised in 1863 to explore the south coast lands.[41] The party, led by Andrew Dempster, sailed eastward from Albany and, after landing at Point Malcolm, travelled along the coast to Point Culver and then inland. In both areas, they found the scattered mosaics of grassland patches and many of the large well-watered granite outcrops described by Roe. The region was deemed ideal and ready for sheep and cattle grazing.[29] Between them, the brothers eventually claimed 1.5 million acres (607 290 ha), from Stokes Inlet to Esperance Bay, including up to 20 Recherche islands close to the coast. As the leases were largely unfenced, the area exploited would have been very much larger. During December 1864, about 518 wool sheep, 80 dual-purpose cattle (such as dairy shorthorns and Herefords) and 90 horses were overlanded to Esperance.[42,ii] The operations established by 1865 at Esperance Bay were constrained by distance to markets, so sheep were primarily for wool, and dairy cattle for butter production as well as meat and milk for local consumption.[29] These non-perishables could be stored and survive long-distance transport. Each year a schooner was chartered to bring in supplies and take the wool clip to market in Adelaide. In 1881, it was noted that sheep for the butcher were overlanded back to Perth.[43] A letter from relative Annie Gull to the Dempsters' aunt Julia Barker related that:

> [after another lambing] ... there are about 1500 from the 600 they started with ... Edward says the cattle are doing well, plenty of calves. It is so cool down there they can make butter all year round and someone in Albany is to take it all from them at the market price.[29]

The homestead at Esperance Bay was described as having good land for a field, a garden and plenty of water. Sheep were pastured around Esperance and Stokes Inlet to the west and further west at Fanny's Cove. Some of the islands were leased and sheep periodically boated to islands including Middle, Cull and Woody Islands. By 1875 sheep numbers had risen to 13 750 and they were being overlanded to Fraser Range and Southern Hills, around 240 km north of Esperance Bay.[44]

Transhumance and mobile shepherding

To succeed in the Esperance bioregion, pastoralists found they needed to adopt management systems that allowed the flexibility to move flocks and herds around at various times of the year. Traditional European sheep management systems included mobile shepherding, in which sheep were constantly moved from grazed to ungrazed areas throughout the year. The system continues in France today, and allows long periods of pasture rest that is especially important to perennial grasses and shrubs.[45] The other was a Mediterranean system called transhumance, originally described by Fernand Braudel, where livestock were shepherded inland to the mountains and back every year. It improved production from sheep flocks, allowed pasture to rest (and may well have improved mineral uptake, as in Australia), and was an accepted and ancient practice that 'offered fiscal resources which no state could ignore'.[46] Transhumance that followed patterns of climatic gradient and markets was integral to any sustainable systems of early pastoralism in eastern Australia, according to Ryan McAllister and co-researchers.[47]

---

ii Old photographs from the period indicate merinos were the dominant breed, with infusions of dual-purpose sheep such as the Lincoln. Source: Esperance Museum Archives, Sheep.

In south Western Australia and the Esperance sandplain, transhumance was crucial as the ancient soils have naturally very low levels of the macronutrients phosphorus, potassium and nitrogen and micronutrients such as cobalt, copper and selenium.[48] Native animals evolved with these low soil nutrients and marsupial grazers, though having ruminant-like plant digestion, require far less cobalt and copper compared to sheep (quokkas need at least 50% less).[49] Ruminant mammals use symbiotic microorganisms for the fermentation in the rumen and hindgut needed to break down plant cellulose. These microorganisms need trace elements, especially cobalt as the vitamin B12 precursor, that animals need for growth and reproduction. Without enough trace elements domestic ruminants fail to grow, become emaciated and die. Horses are hindgut fermenters and, like marsupials, far less susceptible to these trace element deficiencies. The earlier settled west coast colonists had found that for ruminant stock to thrive they needed to spend part of the year inland on heavier soils away from the coast. According to F.A.R. Dempster, Andrew Dempster had experience of this 'coastiness', and knew that the only cure was to take stock inland for a period each year. Very early on, he had looked for suitable inland grazing land to combine with their coastal lands. In 1866 Fraser Range and Southern Hills, about 240 km north-east of Esperance Bay, were decided upon.[50] Time spent on offshore Recherche Archipelago islands was an alternative preventative measure, so those with fresh water and close to the mainland were leased.[29] Middle, Cull and Woody Islands, fertilised by guano from the short-tailed shearwater (*Puffinus tenuirostris*) which nests all over the archipelago, have fertile and mineral-rich soils compared to mainland coastal soils. Livestock grazed on them could be fattened within six to seven months.[29]

Thus, the Dempsters' pastoral enterprises, and later those belonging to Campbell Taylor and the Dimer family, were based on an annual transhumance from the summer coastal pastures to inland winter pastures and to islands. Their ability to run large numbers of stock on inland pastures was further improved by sinking wells around the base of granite outcrops, damming water runoff from outcrops and enlarging existing Aboriginal *gnammas*.[24]

Peter Spooner and co-workers have historical evidence that supports the use of Aboriginal traditional pathways by pastoralists as travelling stock routes (TSRs). The adoption of such pathways into a TSR system likely occurred through Aboriginal guides and trackers passing on existing pathway knowledge. European settlers adopted tracks from observations of physical evidence, and shared development of some TSRs from Aboriginal people working in the pastoral industry. Though this has not been explored in depth for the Esperance bioregion, it is likely that the Dempster Track and other tracks developed for transhumance had Aboriginal origins, as Aboriginal guides and shepherds were commonly included in explorations.[51] The Dempsters' transhumance track was set up so that each nightly stopover had both feed and water, as shown in Figure 6.1.

The ecological benefits of this annual movement of stock inadvertently provided a counterbalance to the erupting sheep herds' potential to cause significant environmental destruction. Its ongoing success was further enhanced by including mobile Aboriginal shepherding in the model. Aboriginal people's grazing management skills, previously applied

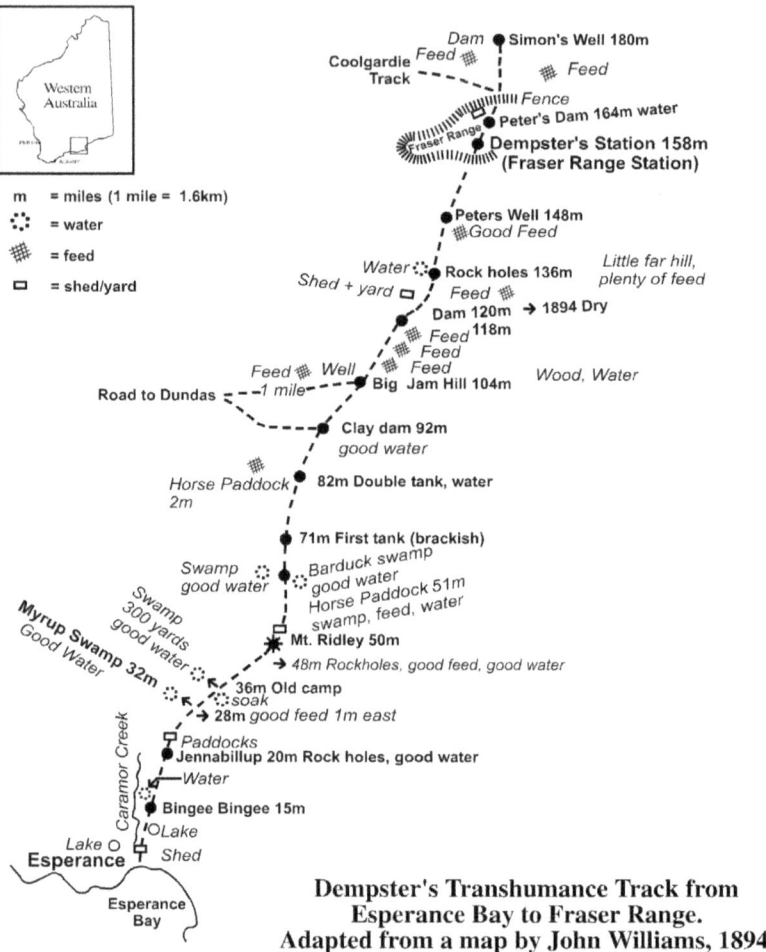

Fig. 6.1: The Dempsters' transhumance track showing the route from Esperance to the Dempster brothers' holdings at Fraser Range. Water, feed and yards are marked. The original transhumance tracks were likely to have been Aboriginal pathways.[52]

to kangaroos, maximised sheep survival and production both coastal and inland, as they used fire to produce fresh grasses and moved sheep around the summer and winter pastures, within the larger annual transhumance pattern.

Aboriginal mobile shepherding: delaying the crash

In 1835 G.F. Moore recorded in his diary, 'the colony is now greatly in want of a few good practical shepherds … [and] It is surprising how much the condition of the flock depends on the goodness of the shepherd'.[53] Shepherding was a highly skilled job – shepherds needed to know how to manage sheep from lambing to shearing, and be familiar with local conditions. As described for Kimberley Tallbear's 'American Indians', for Aboriginal people to survive in the new monetary economy they had to take on aspects of the extractivist practices of the invading colonial consumer front. Shepherding allowed them to do this while keeping connection to Country.[54] Rica Ericson notes that sheep flocks managed by Aboriginal people did very well, and attributed this to Aboriginal shepherds allowing the sheep to range more freely since they could use their tracking abilities to find them quickly for night penning.[43] It is likely that the reasons were more complex than this. They could be explained as eco-shepherding, where fire was used to regenerate pastures and sheep were moved constantly to prevent overgrazing.[55] With their long cultural experience and skills in managing kangaroos, wallabies and other grazers, many Aboriginal people became shepherds to survive the new order. They were documented by Karl Dimer as highly skilled, using traditional (ecological) knowledge about location of the best grasses, poisonous plants to avoid, how and where to follow the thunderstorm rain that produced grass during dry years, when and where dingoes could be a problem, and when native grasses and herbage needed a rest from grazing. Aboriginal shepherds also managed Country with fire to regenerate grasses and keep scrub under control for sheep as they had done for kangaroos and other native animals previously. In a letter to the *Eastern Districts Chronicle* in 1891, a 'Lady' (second-generation settler Amy Baesjou) describes encountering some local Aboriginal shepherds during her ride to Israelite Bay, who were using fire as a shepherding management tool to promote native grasses.[56]

This experienced use of fire is likely to have contributed to the early sustainability of the Esperance bioregion pastoral system. When grazing country is not overstocked there is sufficient grass residue to carry the fire needed to keep unpalatable woody weeds under control and to regenerate palatable grasses and herbs. These then need to be rested to allow root reserves to recover and sufficient seedlings to germinate.[21] Without the Aboriginal people's depth of knowledge about land, animal and ecosystem management which prevented the complete breakdown of existing ecosystems, it is unlikely that pastoralism could have succeeded so well.

Ending pastoralism

The pastoral management systems developed in the Esperance bioregion involved grazing management that initially exhibited a high level of resilience and adaptation to the local conditions. This once adaptive SES declined rapidly in the 1890s and early years of the 20th century, when several coinciding factors led to its demise. It continued for a while in the isolated eastern parts of the Esperance bioregion, but today no longer contributes to sustainability and resilience in Western Australian and other Australian rangelands.

As the Dempsters prospered, economic pressures and the temptation to make more money and increase income seemed to build. Sheep numbers increased until around 35 000 sheep were shorn in the 1890s. This sheep eruption, combined with those of other pastoralists, moved beyond a sustainable threshold regardless of transhumance and eco-shepherding. Fraser Range and Southern Hills in the arid zone were especially impacted by overgrazing. They were unable to recover quickly and drought became more frequent as sheep and other livestock eliminated the perennial pastures. As ex-pastoralist Geoff Grewar explained to me from his personal experience, the now dominant annual vegetation assemblages in rangelands made droughts more frequent and severe, and dramatically lowered the livestock carrying capacity.[57]

During the last years of the 19th century, changes in government policy encouraged closer settlement throughout the Swan River colony to produce more agricultural revenue and help achieve food and grain self-sufficiency. In further regions including the municipality of Esperance (denoted in 1895), yeoman farms developed from the 1890s. With consequent land subdivision and fencing into smaller more intensive farms on fertile mosaics, pastoralists lost their best sandplain and mallee country. The once influential Dempster pastoral company was dissolved in the early 20th century, and the coastal Esperance Bay Station and inland Fraser Range Station were separated. F.A.R. Dempster relates how William Dempster had 'prophesied ruin at Esperance as the separation of the two stations was a grave error because Esperance stock needed the change of herbage to keep them in good condition'.[41,58] Ending transhumance meant large numbers of stock could no longer be kept on the trace element-deficient sandplain. They became confined to farms on patches of fertile soil, such as those found along the Dalyup River.[59]

In the eastern district, pastoralism continued for a period on Nanambinia Station (owned by the part-Aboriginal and German Dimer family). However, productivity declined with the loss of eco-shepherding and cheap labour when Aboriginal people were moved into missions. This led to widespread overgrazing. Pastoralism declined further after 1930, influenced by drying seasons and drought, which Karl Dimer attributed to clearing of the Salmon Gums district 106 km north of Esperance. Though fencing was being used by then, its usefulness was reduced by wild camel damage and the lack of income to employ labour to maintain fences and rotate sheep between paddocks.[57] Without Aboriginal shepherds, a low labour set stocking system became the norm, leading to overgrazing and permanent rangeland damage.

The role of Aboriginal people in the sheep eruption, whether compliant, complicit or compelled by the need for survival in a changed world, was crucial to the role of transhumance and eco-shepherding in mitigating pastoralism impacts. They delayed the destructive effects of overgrazing on soils in the Esperance bioregion, that writers such as Eric Rolls, Neil Barr and John Cary describe as characterising most rangeland pastoralism in western and eastern Australia.[60, 61] The fact that such systems worked so well in Western Australia, and have persisted for thousands of years in France and the Mediterranean world, would surely point to a more sustainable and resilient way of managing Australia's pastoral rangelands.

Endnotes

1. Eyre EJ (1845) *An Account of the Manners and Customs of the Aborigines and the State of Their Relations with Europeans.* T. and W. Boone, London.
2. Crosby AW (1989) Ecological imperialism: the overseas migration of western Europeans as a biological phenomenon. In *The Ends of the Earth: Perspectives on Modern Environmental History.* (Ed. D Worster) pp. 103–117. Cambridge University Press, Cambridge.
3. Melville E (1997) *A Plague of Sheep: Environmental Consequences of the Conquest of Mexico.* Cambridge University Press, Cambridge.
4. Stuer APL (1982) *The French in Australia.* Australian National University Press, Canberra.
5. Porter R (1990) *English Society in the Eighteenth Century.* Penguin, London.
6. Burnett J (1989) *Plenty and Want: A Social History of Food in England from 1815 to the Present Day.* 3rd edn. Methuen, London.
7. Fairlie S (2009) A short history of enclosure in Britain. *Land (Basel)* 7, 16–31.
8. Blaut JM (2012) *The Colonizer's Model of the World: Geographical Diffusionism and Eurocentric History.* Guilford Press, New York.

9. Frawley K (1994) Evolving visions: environmental management and nature conservation in Australia. In *Australian Environmental History: Essays and Cases*. (Ed. S Dovers). Oxford University Press, Oxford.
10. Hunter AA (2012) A different kind of subject: colonial law in Aboriginal–European relations in nineteenth century Western Australia 1829–61. *Journal of Legal History* **14**, 160–164.
11. Blanton DB (2003) The weather is fine, wish you were here, because I'm the last one alive: 'learning' the environment in the English New World colonies. In *Colonization of Unfamiliar Landscapes: The Archaeology of Adaptation*. (Eds M Rockman, J Steele). Psychology Press, London.
12. McGregor R (1997) *Imagined Destinies: Aboriginal Australians and the Doomed Race Theory, 1880–1939*. Melbourne University Press, Melbourne.
13. Wilk R (2004) The extractive economy: an early phase of the globalisation of diet. *Review – Fernand Braudel Center* **27**(4), 285–306.
14. Darwin C, Wallace AR (1858) On the variation of organic beings in a state of nature; on the natural means of selection; on the comparison of domestic races and true species. In *Work on Species*. (Ed. C Darwin). Unpublished, London.
15. Haeckel EH (1899) *On Our Present Knowledge of the Origin of Man*. Publisher unknown.
16. Boom K, Ben-Ami D, Croft DB, Cushing N, Ramp D, Boronyak L (2012) 'Pest' and resource: a legal history of Australia's kangaroos. *Animal Studies Journal* **1**(1), 17–40.
17. Hrdina FC (1997) Marsupial destruction in Queensland 1877–1930. *Australian Zoologist* **30**(3), 272–286. doi:10.7882/AZ.1997.003
18. Anon (1892) Preservation of native game. *Western Mail*, 18 June, p. 38.
19. White S (2012) British colonialism, Australian nationalism and the law: hierarchies of wild animal protection. *Monash University Law Review* **39**, 452–472.
20. Davidson-Hunt I, Berkes F (2003) Nature and society through the lens of resilience: towards a human–in–ecosystem perspective. In *Navigating Social-Ecological Systems: Building Resilience for Complexity and Change*. (Eds F Berkes, J Colding, C Folke). Cambridge University Press, Cambridge.
21. Prober SM, Yuen E, O'Connor M, Schultz L (2013) *Ngadju Kala: Ngadju Fire Knowledge and Contemporary Fire Management in the Great Western Woodlands*. CSIRO Ecosystem Sciences, Perth.
22. Hassell E, Davison DS (1936) Notes on the ethnology of the Wheelman tribe of south-western Australia. *ANTHROPOS: International Review of Anthropology and Linguistics* **31**, 679–711.
23. Colding J, Folke C (2001) Social taboos: 'invisible' systems of local resource management and biological conservation. *Ecological Applications* **11**(2), 584–586.
24. Roe JS (1849) Report on an expedition to the south-eastward of Perth, in Western Australia, between the months of September 1848, and February 1849 under the Surveyor-General Mr John Septimus Roe. *Journal of the Royal Geographical Society of London* **XXII**, 14–20.
25. Moore P (2005) *A Guide to Plants of Inland Australia*. Reed New Holland, Sydney.
26. Caughly G (1981) Comments on natural regulation (what constitutes a real wilderness?). *Wildlife Society Bulletin* **9**, 232–234.
27. Brandis T (2008) *Rescuing the Rangelands: Management Strategies for Restoration and Conservation of the Natural heritage of the Western Australian Rangelands after 150 years of Pastoralism*. WA Department of Environment and Conservation, Perth.
28. Mitchell TL (1848) *Journal of an Expedition into the Interior of Tropical Australia, in Search of a Route from Sydney to the Gulf of Carpentaria*. Longman, Brown, Green and Longmans, London.
29. Ericson R (1978) *The Dempsters*. University of Western Australia Press, Perth.
30. Dominy M (2003) Hearing grass, thinking grass: postcolonialism and ecology in Aotearoa-New Zealand. In *Disputed Territories: Land, Culture and Identity in Settler Societies*. (Eds D Trigger, G Griffiths) pp. 53–80. Hong Kong University Press, Hong Kong.
31. Mitchell PB (1991) Historical perspectives on some vegetation and soil changes in semi-arid New South Wales. *Vegetatio* **91**, 169–182. doi:10.1007/BF00036055
32. Drummond J (1842) Correspondence. *Inquirer* (Perth: 1840–1855), 27 July, p. 4.
33. Sarre A (1999) Slow change on the range. *Ecos* **100**, 44.
34. Vega E, Montana C (2011) Effects of overgrazing and rainfall variability on the dynamics of semiarid banded vegetation patterns: a simulation study with cellular automata. *Journal of Arid Environments* **75**(1), 70–77. doi:10.1016/j.jaridenv.2010.08.001

35. Flannery T (2003) *Beautiful Lies: Population and Environment in Australia*. Black Inc., Melbourne.
36. Gott B (1982) Ecology of root use by the Aborigines of southern Australia. *Archaeology in Oceania* **17**(1), 59–60.
37. Koocheki A, Gliessman SR (2005) Pastoral nomadism: a sustainable system for grazing land management in arid areas. *Journal of Sustainable Agriculture* **25**(4), 113–131. doi:10.1300/J064v25n04_09
38. McDonell N (2016) *The Civilization of Perpetual Movement: Nomads in the Modern World*. 1st edn. C. Hurst & Co., London.
39. Favenc E (1888) *The History of Australian Exploration from 1788 to 1888: Compiled from State Documents, Private Papers, and the Most Authentic Sources of Information. Issued Under the Auspices of the Governments of the Australian Colonies*. Turner and Henderson, Sydney.
40. Brooker L (2006) *Expedition Eastward from Northam by the Dempster Brothers, Clarkson, Harper and Corell, July–August 1861*. Hesperian Press, Perth.
41. Dempster FAR (n.d.) *Andrew Dempster: Founder of Esperance and Muresk*. Self-published, Perth.
42. Tonts M, Yarwood R, Jones R (2010) Global geographies of innovation diffusion: the case of the Australian cattle industry. *Geographical Journal* **176**(1), 90–104. doi:10.1111/j.1475-4959.2009.00331.x
43. Rodgers B (2001) *Transcription of Notes used by Mrs Rica Erickson for her book 'The Dempsters'*. Esperance Municipal Museum, Esperance.
44. Ford F (n.d.) The Country Scene. *The Countryman*.
45. Meuret M, Provenza F (Eds) (2014) *The Art and Science of Shepherding: Tapping the Wisdom of French Herders*. ACRES USA, Austin.
46. Braudel F (1972) *The Mediterranean and the Mediterranean World in the Age of Philip II*. Collins, London.
47. McAllister RRJ, Abel N, Stokes CJ, Gordon IJ (2006) Australian pastoralists in time and space: the evolution of a complex adaptive system. *Ecology and Society* **11**(2), 41. doi:10.5751/ES-01875-110241
48. WA Department of Agriculture and Food, GRDC (2009) *Managing South Coast Sandplain Soils for Yield and Profit*. Bulletin 4773, October.
49. Moir RJ, Somers M, Waring H (1955) Studies on marsupial nutrition 1. Ruminant-like digestion in a herbivorous marsupial (*Setonix brachyurus* Quoy & Gaimard). *Australian Journal of Biological Sciences* **9**, 293–304.
50. Esperance Museum Archives (n.d.) Pioneering Fraser Range. *The Western Mail*. Accession no. 2066. Esperance Municipal Museum, Esperance.
51. Spooner PG, Firman M, Yalmambirra (2010) Origins of travelling stock routes. 1. Connections to Indigenous traditional pathways. *Rangeland Journal* **32**, 329–339. doi:10.1071/RJ10009
52. Adapted from a map by John R Williams, 1894. Photocopy in Esperance Museum 2012/2344/1. Unknown source.
53. Ericson R (1974) *Old Toodyay and Newcastle*. Toodyay Shire Council, Toodyay, pp. 62–68. G.F. Moore quoted.
54. TallBear K (2000) Shepard Krech's *The Ecological Indian*: one Indian's perspective. *Ecological Indian Review*, 1–5.
55. Dimer K (1989) *Elsewhere Fine*. South West Printing and Publishing, Bunbury.
56. Anon (1891) Wood Notes from Western Australia. *Eastern Districts Chronicle* (York, WA: 1877–1927), 31 January.
57. Grewar G (2012) Retired pastoralist. Interviewed by Nicole Chalmer, Coronet Hill, 20 August.
58. Dempster J (1981) Letter to manager of the Esperance Museum concerning the Dempsters.
59. Australian Bureau of Statistics (1911) *Western Australia, Year-Book Australia*. Australian Bureau of Statistics, Canberra.
60. Rolls E (1999) Land of grass: the loss of Australia's grasslands. *Australian Geographical Studies* **37**(3), 197–213. doi:10.1111/1467-8470.00079
61. Barr N, Cary J (1992) *Greening a Brown Land*. Macmillan, Melbourne.

7
Ending Aboriginal social ecological systems and animal landscapes

... very often Indians to survive in the new monetary economy would need to take on the extractivist practices of the invading colonial consumer front.[1]

The invasion of Australia by Anglo pastoralists and their sheep and cattle effectively began the uncoupling of Aboriginal social ecological systems (SES) from their keystone role in co-evolved ecosystems. In Western Australia this was later worsened by the invading rabbit swarms from 1905 onwards. First the sheep and then the rabbits ate the palatable perennial grasses, herbs, yams and tubers that were food for native animals and Aboriginal people. The animal invaders also ate any seedlings that germinated, thus preventing habitat regeneration. In this way animal habitats and food were destroyed, along with much Aboriginal knowledge about traditional foods and the production systems which had shaped identity, cultural, social and ecological systems.[2] The impact of diseases such as measles and influenza on Elders and educators was so swiftly devastating that much of their ecoagricultural knowledge and education systems were diminished within a generation.

Aboriginal people had no choice except to adapt to the new reality or lose their identity. This adaptation was apparent in many districts, as they often worked for pastoralists and maintained some of their traditional hunting practices in unallocated crown land – in effect integrating aspects of their SES into that of the pastoralists. This process could have continued in a positive manner, as I will show in a case study of the Dimer pastoral enterprise in the Esperance bioregion, until it was stifled by the late 19th century encouragement of closer settlement, farm intensification and discriminatory policies towards Aboriginal people.[3]

As a great supporter of the vision of yeomanry and a state economy supported by agriculture, the first Premier of Western Australia, Sir John Forrest, introduced various legislative means to ensure further social and ecological change in the state after self-government was granted in 1891, until the start of World War I. He was also linked to various discriminatory Acts, including further land dispossession and state control of Aboriginal people.[4,5] These Acts removed Aboriginal people's rights and their ability to live on an equitable social footing with the wider Anglo settler society, and were likely influenced by Forrest's drive to increase agricultural productivity and food self-sufficiency in Western Australia. Forrest firmly believed that Aboriginal people were dying out and that 'the Government should do just enough to smooth their passing and to ensure they would fully serve the higher civilisation before they went'.[4,6]

Claiming the land: upheld with law

The 1883 passing of the *Dog Act* in Western Australia to control sheep also allowed the colonial-British legal system to deeply influence and control the lives of Aboriginal Australians. The Act made it illegal for Aboriginal people to keep 'dogs', including dingoes.[7] This effectively reduced their ability to hunt or generate income, as dingoes were integral to Aboriginal hunting systems for both small and large animals.

The introduction of Aboriginal Protection Acts between 1869 and 1910 throughout Australia allowed states and territories to directly impact Aboriginal people's daily lives.[8] The equivalent Western Australian *Aborigines Act* was introduced in 1905 and amended through the *Native Administration Act 1936*.[9,10] Food rationing was one of the government-sanctioned outcomes, along with other controlling measures that regulated where Aboriginal people could live and work, whom they could marry (mixed race marriage was prohibited), where their children could live and what payments they could receive. It was, in effect, a form of apartheid.

These Acts made adaptation to colonialism in a dignified manner by Aboriginal peoples almost impossible.[5] Biological Aboriginality was considered more important than lifestyle and social identification, and anyone with more than one-quarter Aboriginal 'blood' was brought under the Act regardless of how they identified themselves. It severely limited their ability to work, and their freedom to choose where to live and with whom. They were no longer the legal guardians of their children (the Chief Protector of Aborigines was), could be ordered out of towns or moved to reserves, and it was illegal for any non-Aboriginal person to associate with or marry an Aboriginal person. 'Certificates of Exemption' that allowed greater freedom of movement and employment could be obtained with difficulty: the recipient would 'cease to be an Aborigine' under the Act. Such a certificate could be revoked at any time by the responsible Minister.[5] In order to maintain their freedoms, many people went to great lengths to disguise their Aboriginal identity, creating a barrier to maintenance of culture and traditional land management practices.

Population collapse: disease and hunger

Throughout Australia, Anglo settlements and towns took over the most desirable areas of productive lands – where Aboriginal populations were densest. They were then accused of 'hanging about' towns in 'undesirable ways'! Many Aboriginal people were reported as living on their traditional lands in the Esperance bioregion in the early colonial days. Amy Crocker, a descendent of the Pontons who took up land east of Esperance in 1873, makes it clear in her diaries that there were numerous Aboriginal people living in the area around Mt Ragged.[11] By the early 1900s, after 40 years of pastoralism, it is doubtful that the bioregion would have supported this original population without outside food inputs. These areas were rapidly emptied as Aboriginal groups died of introduced disease or were displaced. The land's carrying capacity for a society reliant on European livestock and plants was lower than that for an Aboriginal society that relied on the local native plants and animals.[12]

Deaths and infertility from epidemics of novel diseases such as measles, influenza, smallpox and venereal diseases are well recognised as the major cause of Aboriginal depopulation throughout Australia.[13] Less recognised is that chronic starvation, enforced crowding into reserves and dramatic dietary changes would have also caused mortalities and greatly increased susceptibility to diseases. These physical challenges would have been magnified by the psychological impacts of intense solastalgia. This concept, developed by philosopher Glen Albrecht and partners, refers to the intense distress and depression that occurs when people are exposed to environmental change 'that is exacerbated by a sense of powerlessness or lack of control over the unfolding change process'.[14] The forced removal of Aboriginal people from their traditional lands, their basis for all existence, is reported by Daisy Bates to have left some in a catatonic state.[15]

As Anglo pastoralism spread and consolidated, causing Aboriginal ecoagriculture to collapse, hunger become pervasive as not enough nutritious native foods could be produced. Anglo food substitutes of far lower nutritional value, such as flour, tea and sugar, became dietary staples as food rationing became established, initiating a cycle of semi-starvation and poor health.[16] Food is not only a complex combination of nutrients to feed the body and brain, but a social process vital to a functioning culture and landscape belonging. Sebastian and Donnelly consider that the best way to disempower and undermine a SES is to remove the ability of people to access their traditional foods. Carter and Nutter confirm this when they describe how quickly Nyungar people ran out of both plant and animal food as it was eaten or destroyed by the Anglos of the Swan River settlement.[8,17] By deliberately interfering with participatory, well integrated, complex and resilient food systems, rationing also contributed to the breakdown of cultural practices.

In Western Australia, the government rations that replaced traditional foods consisted of a yearly allowance per person of 20 lb [9 kg] flour (or meat to equivalent monetary value), 3 lb [1.3 kg] sugar and 0.5 lb [250 g] tea. These were supposed to be supplemented with the rapidly disappearing bush tucker, and help from settlers. This was not a balanced diet for a people with a dietary history rich in protein, roughage and nutrient-dense plant foods obtained from a wide variety of sources. Such foods, high in refined carbohydrates and sugars, are now recognised as deadly to people such as Aboriginal peoples, who are genetically insulin-resistant.[18] The meagre and toxic ration was then halved by Premier John Forrest in 1898, due to his conviction that the ration system had made Aboriginal people 'lazy'.[5] The impacts may have been even greater if not for the arrival of rabbits in 1901, which became a staple food for many Aboriginal families.[2] They replaced the greatly diminished or locally extinct populations of small marsupials as a food source.

Before the 1850s the voyage from Britain to colonial Western Australia was about 100 days, serving as an efficient form of quarantine. Those who became sick onboard either died or recovered and were no longer infectious by the journey's end. During this period, European colonists and Aboriginal peoples alike were generally healthy and disease-free – with no records of Aboriginal people dying of introduced diseases. The fatalities began with the steam ship and clipper revolution from the 1860s to 1880s, which reduced the journey to 50–80 days and ended the quarantine effect.[19,20] A measles outbreak was brought by ship to Albany in 1860, and the disease became established in the colony. Aboriginal people were particularly vulnerable due to their lack of immunity and their generally poor health, caused by semi-starvation, a nutrient-deficient diet and solastalgia. By 1883 severe measles outbreaks were recorded by Emily Dempster who, along with her husband Andrew, nursed many sick people and reported many Aboriginal deaths around Esperance Bay.[21] According to Jesse E. Hammond, this outbreak throughout south-western Australia had devastating effects on the local Aboriginal populations:

> The measles epidemic of the early 'eighties' affected the natives of the South-West and East very much. They died off in great numbers; and the nature of those that were left was altered; they lost all interest in bush life; they did not care what the others did or where they went and they were never the same people again. They dropped their own tongue and used the white man's language; they drifted away from all laws, ceremonies and customs.[22]

Hammond's claim that the nature of survivors was permanently altered by the impacts of measles outbreaks should be understood as part of a raft of coinciding physical and emotional shocks instigated knowingly and unknowingly by the invading colonists. The intense grief at loss of loved ones, combined with ongoing alienation and loss of Country that people rightfully regarded as their own, would also have intense depressive impacts and solastalgia. Elders, both female and male, were especially vulnerable. As they were solely responsible for holding and disseminating cultural knowledge and education, their untimely deaths and loss of status would have made retention, continuation and dissemination of the ecological knowledge to manage a complex SES very difficult for survivors.[23,24] In Esperance, local police estimated that by 1900 there were 300 Aboriginal people in the district compared to 420 Europeans and by 1912 the Aboriginal population was reduced to 140.[25] This corresponds with Daisy Bates' deduction that within 40 years of European contact, influenza, measles and STDs had reduced the south-western Australian Aboriginal population by 80%.[15]

The Spanish flu outbreak in 1917 was a further blow, particularly to the elderly and young people. Older people who could no longer work for sufficient food rations on stations were especially undernourished. In the Esperance bioregion, undernourishment continued into the 1920s–1930s for, as Tom Dimer described 'there was no nourishment in the Government rations for sick people, as it was only flour, sugar and tea'. He described how his family supplied mutton and other nutritious food to some of the camps. The cold weather led to increased deaths in Norseman, which prompted his father Henry Dimer to write a letter to the Chief Protector of Aborigines in 1932 saying that 'the Aborigines would be extinct' if they were not issued good-quality food rations.[3,26]

Aboriginal people in the far eastern Esperance bioregion were encouraged by the state government to move off their Country to the Norseman Mission. Once there, boredom, alcohol and the poor diet of government rations further impacted their health and well-being. Tom Dimer tells:

> The Aborigines lived to a great old age in the bush at Nanambinia … when the mission was started in Norseman they congregated in Norseman and killed themselves with booze … neglected themselves – get dole and go and buy some beer.[26]

Resistance and conflict

Within 20 years of pastoralism becoming established in the Esperance bioregion, Aboriginal people responded by stealing food and killing sheep.[27] Neville Green's research into Aboriginal prosecutions in Western Australia, spanning the period 1831–1898, explains how during the 1880s and 1890s, the primary reason for south coastal Aboriginal people to appear before the Esperance and Albany courts was sheep stealing. He cites repeated records of Aboriginal people arrested for this, frequently at Esperance and Fraser Range by the Dempsters, tried in Albany and sent to Rottnest Island to serve sentences.[23]

Though massacres have not been considered significant in Aboriginal population decline, the evidence of their shameful impacts at local levels is gradually being acknowledged. There was a massacre of Aboriginal people at Cocanarup near Ravensthorpe, related to rape as well as sheep stealing. Cocanarup was a property taken up by the Dunn brothers in 1872 as a sheep run. In 1880 John Dunn was speared and killed by an Aboriginal man (Yandawalla) for his part in raping a 13-year-old Nyungar girl. It seems the Dunn brothers received a government permit to kill the 17 local people who were dubbed the instigators; eventually over 30 people were killed, with survivors fleeing and abandoning the region.[28]

As pastoralists and their sheep flocks encroached further and the numbers of food animals and plants declined, there were outbreaks of conflict as Aboriginal people attempted to protect their Country and way of life. At Fraser Range, Aboriginal people made it clear that the Dempsters were not welcome. A diary note dated Sunday 7 June 1870 described how water infrastructure had been destroyed and 'they have also burned down all our huts'.[29] It is reasonable to assume that the impacts of pastoralism and sheep had been well communicated between Aboriginal groups throughout the region and perhaps even wider, as illustrated by the immediate resistance at Fraser Range. Gifford states that the early police station in Esperance was seen as necessary because of trouble from Aboriginal people in the district.[25] In 1879, there were several demonstrations by Aboriginal people in Esperance township against the practices of European settlers. Police officer John Malcolm described how 'it was found necessary to segregate the ringleaders', who were imprisoned on Black Island which then became regularly used for this purpose.[30]

As Karl Dimer pointed out in 1989, 'Some Aborigines did kill a few sheep, but who can blame them? The sheep ate the grass that their game used to live on'.[3] Not only did the sheep eat grass and other herbage but, as John Pickard describes, an Australia-wide problem was that pastoralists had no idea what levels of stocking the land could sustainably carry. The inland climate at Fraser Range could be extremely variable, so land which appeared understocked in some years would be overstocked in others. Pickard contends that even in modern times there can be no set rules for carrying capacity in most of Australia's rangelands.[31] In the early days these problems were overcome by occupying more land, the practice of mobility through transhumance and the use of mobile strategic shepherding. The Dempsters eventually expanded their original lease areas and occupied the best coastal lands from Hopetoun (then Maryanne Haven) to east of Esperance. Inland they occupied areas around Fraser Range that enabled transhumance, alienating more and more Aboriginal Country and devastating food ecosystems and habitats on a regional scale with their erupting sheep numbers.[32]

The regional destruction of ecosystems, local massacres and disease impacts meant that Aboriginal people's resistance could not be sustained. Survival meant acceptance of European occupation of Country, and adaptation to the situation. An example of how this could have been positive on a large scale, in terms of integrating two SES, is given by the Dimer family, who were descended from a German father and mixed descent Nyungar Aboriginal mother. The family owned the lease of Nanambinia Station in the eastern Esperance bioregion.[3] Their version of colonial mobile pastoralism included Aboriginal people as part of an enterprise that integrated pastoralism and family-scale agrarian practices with Aboriginal land management practices. In the early days, this German–Aboriginal family also displayed a strong attachment to their Country and detailed knowledge of how to best manage it.

Nanambinia Station: a template for pastoral sustainability?

Nanambinia Station was a unique community of full- and mixed-descent Aboriginal people who lived together, with individual roles contributing to its overall success. Aboriginal people provided low-paid labour, and the family members worked just as hard for little individual income. As the youngest son Tom Dimer explained in 1990:

> … all the stations round there had Aboriginal shepherds you see … all they wanted was clothes, tucker and tobacco. And it's cheap labour, cheap labour them days. And that's how all them stations started off.[26]

Heinrich (Henry) Dimer, a German butcher turned sailor, jumped ship in Albany in 1884 and worked for Campbell Taylor at Thomas River Station east of Esperance. He worked for other early settlers as well, building his flock by taking sheep in payment. In 1901 Henry was granted an 8094 ha (20 000 acre) pastoral lease north-east of Point Culver and in 1902 was granted a further 10 522 ha (26 000 acres). This became Nanambinia Station (after the Aboriginal words *narnoo*, a small local wattle and *binya*, small).[3] With a further lease east of Nanambinia acquired in 1903, the station was of a viable size. At Nanambinia Station, Henry built a home for his wife Topsy (who was of Aboriginal descent) and their many children. There seemed little distinction in terms of gendered labour, as both boys and the girls helped with station and domestic work. The keenest hunter was Annie, who 'was a great one for hunting anything, from white ant nests to rabbits, turkeys, goannas and their eggs'. The men in the family were accomplished cooks and fruit and vegetable bottlers.

The station is recorded as having abundant herbage for grazing domestic animals including saltbush varieties, blue bush, sage bush, samphire bush, grey broom and quandong trees. There were plentiful perennial grasses and low herbage, especially after rain. Rangelands expert Jim Addison believes changed fire regimes, rabbits and overgrazing have led to annual grasses dominating the once rich and productive chenopod shrublands that originally featured in this area.[33]

Water was the ongoing concern as there were no permanent water sources for large numbers of domestic stock. Pastoralism's first fundamental ecological impact in arid regions was the provision of permanent water with dams or wells, allowing livestock carrying capacity to be increased but resulting in overstocking, especially around water sources, by

Fig. 7.1: Digging out Salmon Gums railway dam at Salmon Gums township, c.1927.[34]

wild as well as domestic herbivores. Using camel- or horse-pulled scoops (see Fig. 7.1), the Dimers built several large dams and improved catchments to increase the amount and permanence of water.

However, erratic rainfall, especially from the 1930s to 1940s, meant that water remained a limiting factor. Karl Dimer noted 'the scarcity of water, in fact, tended to rule our lives'.[3] Water scarcity and continual shepherding prevented overstocking, until new government lease conditions for pastoral land were introduced in the 1930s. These conditions reflected a policy aimed at rapid settlement of the most productive and accessible land, but they ensured overstocking.[35] Lease rates and liability for forfeiture were linked to the average price of greasy wool produced in Western Australia as declared by the government statistician, not to rangeland carrying capacity, seasonal conditions or drought. The facts that rangelands were not consistently productive year in year out, and that such prescriptive stocking rate policies contributed to their ongoing degradation, were completely ignored.[36] Astonishingly, the same minimum conditions still exist today despite the body of research showing them to be completely unsustainable.[37]

Rangeland ecologist Tony Brandis relates how sheep numbers rose following the 1922–1928 wool boom, so that by 1934 the Western Australian rangelands were carrying up to 5.5 million sheep.[35] In reality rangelands have no fixed carrying capacity because, with frequent dry years interspersed with occasional high rainfall events, climatically it is particularly difficult to match forage demand (stocking rate) with supply (available forage).[38] In the 1930s, Francis Ratcliffe was employed by the Commonwealth government to establish the causes of increasing soil degradation in arid South Australia. He determined the causes to be livestock overstocking caused by financial considerations, reluctance to remove stock during drought, destruction of perennials and their seedlings especially in drought, very slow regeneration as long-lived perennial plants cannot rapidly rebound, and increasing drought frequency linked to vegetation destruction.[39] However, nothing much changed despite his report as it is easy to overstock rangelands, causing an environmental disaster heralding desertification.[40]

On Nanambinia the main enterprise was sheep for wool. Shepherding was done expertly by Aboriginal families who lived permanently with their mobs of sheep, receiving food and other material supplies to supplement their low wages. A licence was needed to employ Aboriginal people and their payment in food and other material needs was supplemented by government rations and traditional foods, initially still available. The station also bred horses, some of which were sent to a business partner who broke them to saddle before sale. Others were kept for station work such as pulling carts and sulkies and dam building. Donkeys were kept for pulling carts, ploughing and general duties that required strength rather than speed, and because they were easy for young Dimer children to handle. However, camels became the major power source for station work as they required far less water than horses, could eat a wider range of plant foods and had superior draught abilities (see Fig. 7.2).[3]

Cattle were run at Point Culver and, with plentiful runoff water from the dam at a large granite outcrop, a large vegetable garden and an orchard were established at Nanambinia homestead. The family and employees were largely self-sufficient with a form of ecoagriculture where domesticated plant and animal foods were supplemented with wild food plants (quandong, yams) and animals (kangaroos, wallabies, boodies, bush turkey, goannas, fish and water birds). Patches of land were cleared to grow oats and wheaten hay for the dairy cows and horses. When Henry Dimer found that the cleared saltbush land was too salty to grow a crop beyond two years, he confined crops to the red soils surrounding granite outcrops. It took around 50 years for the cropped saltbush land to poorly regenerate patchy native grasses.[3]

Fig. 7.2: A Dimer family camel team, carting wool to Israelite Bay.[41]

Nanambinia Station continued as a family enterprise after the death of Henry in December 1936. Its success related to multiple and flexible management systems based on a largely unpaid, extensive workforce of family and Aboriginal people, along with integrating Aboriginal systems of managing Country into the station enterprise. Land management was closely monitored yet flexible, keeping stock and land in good condition with constant strategic shepherding movements to new grazing and the resting of grazed areas. Having coastal leases allowed transhumance and scrub patches were burnt in January, in a yearly rotation to regenerate grasses and herbaceous plants. Income risk was spread between different enterprises including wool, beef, horses and kangaroos, with dingo and eagle bounties. Drought impacts were buffered with the higher-rainfall coastal leaseholds, the construction of large dams on inland properties and the ability of Aboriginal shepherds to know when and how to follow rain-producing thunderstorms with the sheep flocks.[3] Food self-sufficiency allowed much of the cash income to be used for improvements or off-station investments (for example, a butcher shop was purchased in Norseman). These factors allowed the business to be fairly resilient to environmental changes and economic conditions.

The successful Dimer family enterprise lasted one and a half generations before pastoral lease changes, social and ecological factors and climatic vagaries impacted it. The Aboriginal workforce disappeared as the numbers of Aboriginal people declined, and they were 'encouraged' to enter Norseman missions during the 1930s. Without Aboriginal shepherds and their ecological skill sets, overgrazing with set stocked sheep and other livestock eliminated drought-resistant perennial pastures, resulting in reliance on ephemeral annual grasses.[33] Pastoralism was increasingly difficult between 1926 and 1947, a 20-year period during which, according to Karl Dimer (supported by Bureau of Meteorology data), the region became significantly drier and subject to droughts.[3] It was likely exacerbated by overgrazing and destruction of the hydrological microclimate. At the same time, large-scale clearing was taking place at Salmon Gums, with possibly adverse impacts on local climatic patterns.[3]

By the 1950s the station was unable to provide a living and it went into a cycle of slow decline. There were income disputes – a disposable cash income was needed if the family were to live in parity with the rest of society, with a car, modern appliances and holidays. The station could not provide this without going into debt for significant expansion. Nanambinia was sold in the 1980s as the family no longer wanted to live in isolation or work so hard, and couldn't afford the necessary labour. Following the 'get big or get out' pattern described by agricultural researcher Neil Barr, today Nanambinia is part of the six-property aggregation of Balladonia Station.[42,43]

Ending the animal–plant landscapes

Destructive ecosystem changes in Australia started with pastoralism impacts that contributed to the loss of Aboriginal SES. Fire was no longer managed, yam beds were no longer dug over, habitats were destroyed, and the lives of important animals and plants were no longer protected.

Information found in diaries, journals and oral histories provides a baseline for the animals present in the Esperance bioregion when the Dempsters first arrived. There were large visible species such as emus, bush turkeys and grey kangaroos living in the Aboriginal-managed patchy grassland habitats. The cryptic animals of the swamps, thickets, trees and edges of clearings and granite outcrops included snakes, lizards, frogs, quolls, bandicoots, woylies, bilbies, tammar wallabies, black gloved wallabies, quokkas, rock wallabies, and brushtail and ringtail possums. In the drier areas to the east (Israelite Bay and north) and north to Fraser Range and Southern Hills, pastoralist Amy Crocker describes burrowing bettong, bilby and wombat warrens scattered throughout the mallee woodlands. The activities of these animal residents complemented Aboriginal burning practices by maintaining grassy patches as they ate regenerating eucalypt and acacia seedlings and fertilised the open areas around their burrows with droppings.[44] Also present were woylies, red kangaroos, euros, emus, bush turkeys, mallee fowl, hare wallabies and western barred bandicoot species.[i] Distributed throughout the micro-ecologies from coast to mallee woodlands were smaller marsupials – honey possums, pygmy possums, and small predatory dasyurid species including antechinus, dibblers and pouched mice. There were small rodents such as hopping mice, stick nest rats, western mice and kangaroo mice as well as numerous species of insects, reptiles and frogs. As hunters and plant food collectors, Aboriginal peoples were at the apex. Below them in the predatory hierarchy were the smaller hunters such as dingoes, eagles and hawks, western quolls, monitor lizards, phascogales and antechinus species. The trees were occupied by ringtail and brushtail possums and birds.[45,46] Flocks of Carnaby's cockatoos in the hundreds of thousands were still a seasonal feature in the 1950s skies, migrating between coastal feeding grounds and nesting areas in the ancient mallee woodlands to the north and west, until major clearing commenced during the 1960s and destroyed nesting trees and food habitat.[47]

Tim Flannery describes how pioneer journals repeatedly record native fauna disappearance throughout Australia following prolonged sheep overgrazing. In South Australia's Flinders Ranges, gross overgrazing from 1860 to 1880, with consequent habitat destruction and loss of the rainfall that relied on native vegetation patterns, is closely correlated with extinction of medium-sized native mammals.[48,49] Burbidge and McKenzie describe how a universal feature of extinct Australian mammals is a critical weight range (CWR) of 3.5 g to 5 kg. They link this phenomenon to loss of the habitats typified by the thick layers of herbs and grasses that these species need for food and shelter.[50] Tom Dimer uses the Ngadju name 'little wirra' (perhaps woylies?) for a small marsupial once common in the mallee woodlands, that wove grasses into a nest for shelter. It had disappeared by the late 1920s.[26] 'Little wirra' would have been very susceptible to any level of overgrazing, especially during drought.

Swamp-growing typha species, with their starchy rhizomes, were an important Aboriginal food and their dense, tall leafy stems provide food, shelter and breeding habitat for small mammals, frogs, water birds and insects. Unfortunately they are very palatable to cattle, which can quickly consume them and destroy wetland habitat. The loss of typha and their freshwater wetlands Australia-wide has caused the drastic decline into rarity of the Australasian bittern (*Botaurus poiciloptilus*).[51,52]

---

i   Euro is the Western Australian name for wallaroo.

Recent studies nominate domestic animal grazing, inappropriate fire regimes and large predator control as the primary factors in triggering native ecosystem collapse. Habitat loss as the critical factor in recent declines of northern native rodents (not cats) has been confirmed by Michael Lawes and co-researchers.[53] Using a multivariate comparative approach, Diana Fisher and colleagues compared extrinsic and intrinsic factors leading to the decline and extinction of marsupials. They concluded that habitat degradation by sheep (or cattle) is the most important factor, which secondarily allows predation pressure.[54] The elimination of dingoes as a keystone predator (the only one now left in Australia) allows the proliferation of smaller predators such as varanid lizards, foxes and cats which can use rabbits and introduced murine species (rats and mice) as a base for persistent native animal predation.[55] If dingo control was reduced, new functional ecosystems could emerge with fewer smaller predators, but still enough to have an important role in controlling rabbits and the vast numbers of introduced rats and mice in Australia. This is an area seriously lacking in research.

Humans and other predators

Historian Thomas Dunlap details how the common approach to indigenous predators in the European diaspora was to get rid of them.[56] In Western Australia, dingoes became a declared pest in 1885 and the Dog Acts of 1883 to 1903 limited the number of dogs an Aboriginal man could have, to one. A bounty system paid 10 shillings to any person killing a wild dog – the tail was required as proof – so making them an attractive income source for Aboriginal people and settlers alike. Karl and Tom Dimer both describe the targeting of dingoes because of their potential and actual predation on sheep and their bounty value, as well as how their numbers drastically declined.[3,26,57] The policies effectively prevented proliferation of feral dogs and hybrids, resulting in the almost complete dominance of *Canis dingo* genes that Danielle Stephens has found in modern populations of Western Australian dingoes.[58] Stephens has developed a molecular gene map of hybridisation between dogs and dingoes across Australia. In the Esperance and Kalgoorlie bioregions and majority of Western Australia, over 95% of the sampled 'feral dogs' were genetically dingoes. The highest percentage of dog–dingo hybridisation is on the highly populated eastern coast of Australia.

Initially dingo predation on sheep would have increased, as did sheep stealing by Aboriginal peoples, when sheep numbers increased significantly, eating out habitat so that native food animals declined. Not only would Aboriginal people have been hungry but so would the smaller predators – dingoes, wedgetail eagles and western quolls. The two larger species were targeted relentlessly from early colonisation, with government bounties paid to encourage their elimination. The rabbit invasion of 1905 provided food that helped Aboriginal people, dingoes and eagles to survive.[59] In the Nullarbor today, wedgetail eagles have a 50% rabbit diet as specialist rabbit hunters. The introduction of the rabbit calicivirus which decimated the rabbit population, caused many hungry eagles to relocate to coastal farms near Esperance (including Coronet Hill) in an attempt to find food.

As human and other predation was removed, there was a period of predator release for some populations of marsupial prey animals.[60] In parts of south-western Australia, smaller marsupial populations increased to the level of causing nuisance to European agricultural settlers, as shown in this letter to a newspaper:

> Probandus. SIR, -Almost every settler in the colony most bitterly feels the rapid increase of the above-named little wretches [rats and pordies] which torment us at seed-time and harvest, and all the year round if they can possibly get at our gardens or anything else … the natives no longer seek their living by killing, kangaroo rats and digging out pordies … we destroy, all the native dogs we can … and likewise the eagles, which altogether natives, dogs, and eagle-hawks …. tended greatly, to keep them under.[61]

The commercial hunting of kangaroos and wallabies, killed in large numbers on the Western Australian south coast, was initially to supply meat to sealers and whalers. In the 1880s, skins and furs became the main commodity. Brushtail possums, an important Aboriginal food resource, were heavily impacted as possum fur was particularly desirable as part of the Anglo/European fashion industry and by 1904 had become Western Australia's fifth most important export item.[3] In areas where prime tree habitat was limited, such as the Esperance bioregion where possums (and now most large trees) are now extinct, it would have been easy to cause local extinction or severe decline in brushtail and ringtail possums. So assiduously were they and other species hunted for profit, that in Western Australia the *Game Act 1892* was introduced in an attempt to protect them. When numbers eventually increased the *Game Act* was amended in 1912–1913 so that government royalties were taken from skin sales, reinforced with a hefty penalty of £50 for non-compliance, to be paid to the state government.[62] Northern sea birds were recognised as the source of guano being extracted as fertiliser from north-west islands, so were protected, as was the state emblem, the black swan (*Cygnus atratus*). It is doubtful that this was enforceable in vast remote regions with minimal police.

Marsupials and disease in Western Australia

South-western Australia has lost far more marsupial species than the rest of Australia. Throughout the world the major role of new diseases in mortalities of naive human populations is well documented, though such impacts on native animals is largely unrecognised.[63] If a disease event was part of a 'perfect storm' concurrent with overgrazing, habitat destruction, human overhunting, a rabbit invasion and fox predation, surviving mammal populations are doomed. Ian Abbott has researched extensive historical data and held interviews with people alive at the time of the mammal die-offs. He proposes that an epizootic disease originated at Shark Bay in the late 1870s–1880s, perhaps from pets (monkeys, civet cats, cats or dogs) or rodents on Malaysian pearling luggers. This disease was a major contributor to the drastic marsupial and native rodent decline in southern and northern Western Australia, in some areas 'even before habitat destruction by sheep'.[64] He proposes that the disease first moved through the rangelands north and south of Shark Bay, reaching the south-east of Western Australia by 1890 and by 1920 it affected the whole south-west and Kimberley, leaving only the deserts disease-free.

Amy Crocker's recollections of Balladonia Station support this theory: 'My uncle told me of a strange virus which attacked all small marsupials, killing many species right out. Possums were among these. I think this occurred in the 1880s or 1890s'.[46] She remembered that 'bilbies and chuditch [native cat]' may have survived the disease because they were still present in 1917 in reasonable numbers until foxes became established. A settler farmer from Denmark west of Albany described that as a child in the 1920s he saw numbers of brushtail possums dead under trees, before they disappeared from the forests.

The decline of small and medium-sized mammals would have profound ramifications for ecosystem processes. Soils would become compacted and eroded. Water infiltration, aeration, carbon sequestration, nutrient recycling by soil microorganisms and seed germination, and persistence of perennial herbage, significantly reduced as the small ground-foraging and digging keystone mammals such as the woylie and burrowing bettong disappeared. Ecologist Greg Martin is convinced that their activities mitigated fire intensity and so, coupled with reduced Aboriginal burning practices, control of fire was lost in large parts of Australia.[65] The habitat complex of open herbaceous and grassed areas produced by burrowing bettong grazing patterns once allowed many other small co-dependent animals to flourish. James Noble and co-researchers describe how the regional extinction of *mitika* (burrowing bettong) populations in Central Australia is now widely recognised as a symptom of desertification.[66]

In all Australian landscapes, flora and fauna co-evolved for so long with Aboriginal SES as the keystone culture that as their role discontinued, along with disease and intense livestock overgrazing, ecosystems began to collapse. This collapse continued to completion as closer settlement, fencing and land clearing expanded after World War II. Farms industrialised, and towns and cities expanded further with urbanisation covering the best food- growing country. Aboriginal SES and their beautiful animal and plant landscapes are now a memory impossible to reinstate.

Endnotes

1. TallBear K (2000) Shepard Krech's *The Ecological Indian*: one Indian's perspective. *Ecological Indian Review*, 1–5.
2. Gifford P (2002) *Black and White and in Between: Arthur Dimer and the Nullarbor*. Hesperian Press, Perth.
3. Dimer K (1989) *Elsewhere Fine*. Southwest Printing and Publishing, Bunbury.
4. Tonts M (2002) State policy and the yeoman ideal: agricultural development in Western Australia, 1890–1914. *Landscape Research* **27**(1), 103–115. doi:10.1080/01426390220110793
5. Haebich A (1998) *For Their Own Good: Aborigines and Government in the South-west of Western Australia 1900–1940*. University of Western Australia Press, Perth.
6. Goddard E, Stannage T (1984) John Forrest and the Aborigines. In *European-Aboriginal Relations in Western Australian History*. (Eds W Reece, T Stannage) pp. 55–56. University of Western Australia Press, Perth.
7. *Dog Act 1883*, Western Australia.
8. Sebastian T, Donnelly M (2013) Policy influences affecting the food practices of Indigenous Australians since colonisation. *Australian Aboriginal Studies* **2013**, 59–75.
9. *Aboriginal Act 1905*, Western Australia.
10. Australian Institute of Aboriginal and Torres Strait Islander Studies (n.d.) *Western Australia Legislation / Key Provisions*. <https://aiatsis.gov.au/collections/collections-online/digitised-collections/remove-and-protect/western-australia>
11. Crocker AE (1986) *To Strive, to Achieve, to Leave a Splendid Memory: Being a Brief History of the Pioneering Pontons and Their Descendants of Balladonia Station*. A.E. Crocker, Esperance.

12. EMA HS/832. J.P. Brooks. *Journal of a Trip to Eucla Undertaken in the Spring of 1874.*
13. Flood J (2006) *The Original Australians: Story of the Aboriginal People.* Allen and Unwin, Sydney.
14. Albrecht G, Santore GM, Connor L, Higginbotham N, Freeman S, Kelly B et al. (2007) Solastalgia: the distress caused by environmental change. *Australasian Psychiatry* **15**(1), 5–98.
15. Bates D (1936) *My Natives and I – Incorporating the Passing of the Aborigines: A Lifetime spent among the Natives of Australia.* (Ed. PJ Bridge). Hesperian Press, Perth.
16. Rowse T (1998) *White Power, White Flour: From Rations to Citizenship in Australia.* Cambridge University Press, Melbourne.
17. Carter B, Nutter L (n.d.) *Nyungah Land: Records of Invasion and Theft of Aboriginal Land on the Swan River 1829–1850.* Black History Series, Swan Valley Nyungah Community, WA.
18. Brand Miller JC, Colagiuri S (1994) The carnivore connection: dietary carbohydrate in the evolution of NIDDM. *Diabetologia* **37**(12), 1280–1286. doi:10.1007/BF00399803
19. Australian National Maritime Museum (n.d.) *Passenger Ships to Australia: A Comparison of Vessels and Journey Times to Australia Between 1788 and 1900.* <http://www.anmm.gov.au/Learn/Library-and-Research/Research-Guides/Passenger-Ships-to-Australia-A-Comparison-of-Vessels-and-Journey-Time Accessed 18/08/2018>
20. Campbell J (2007) Western Australia 1860–1870. In *Invisible Invaders: Smallpox and Other Diseases in Aboriginal Australia, 1780–1880.* (Ed. J Campbell) pp. 194–195. Melbourne University Press, Melbourne.
21. EMA 2029 (1968) E. Cotton. *The Dempsters of Esperance Bay.*
22. Hammond JE (1933) *Winjan's People.* Imperial Printing Co., Perth. Facsimile by Hesperian Press, Perth, 1980.
23. Green N (1984) *Broken Spear: Aboriginals and Europeans in the Southwest of Australia.* Focus Education Services, Perth.
24. Kelly L (2016) *The Memory Code.* Allen and Unwin, Sydney.
25. Gifford P (2002) *Black and White and in Between: Arthur Dimer and the Nullarbor.* Hesperian Press, Perth.
26. Dimer T (1990) *Outback Station Life; Aboriginal Customs and Beliefs; Bush Skills and Survival; Vermin Control for Agricultural Protection Board.* OH 2339 – Interviewer, Helen Crompton (transcription). March. Oral History Unit, J.S. Battye Library, Perth.
27. A letter written by Miss Eliza Dunn (original at Ravensthorpe Historical Society) (2008) *Ravensthorpe Then and Now.* (Ed. AW Archer) pp. 10–14. PK Print, Perth.
28. Forrest RG (2004) *Kukenarup: Two Stories. A Report on Historical Accounts of a Massacre Site at Cocanarup near Ravensthorpe W.A.* Yarramoup Aboriginal Corporation, Department of Indigenous Affairs, Perth.
29. EMA HS/694 DEM. Extract from diary of a Dempster brother, when pioneering Fraser Range, 1870.
30. Malcolm J (1928) Pioneer police early days at Esperance. *WA Sunday Times*, June.
31. Pickard J (2001) Carrying capacity in semi-arid Australia. In *Land Degradation: Papers Selected from Contributions to the 6th Meeting of the International Geographical Union's Commission on Land Degradation and Desertification*, Perth, 20–28 September 1999. (Ed. A Conacher). Kluwer Academic, Dordrecht.
32. Erickson R (1978) *The Dempsters.* University of Western Australia Press, Perth.
33. Addison J (2017) Former Senior Technical Officer, WA Department of Agriculture and Food, Kalgoorlie. Pers. comm.
34. EMA, P474–1. Source unknown.
35. Brandis A (2008) *Rescuing the Rangelands: Management Strategies for Restoration and Conservation of the Natural Heritage of the Western Australian Rangelands after 150 Years of Pastoralism.* WA Department of Environment and Conservation, Perth.
36. *Land Act 1933* Western Australia.
37. Duncan W (2017) Former Western Australian Senator who grew up on a station. She lobbied for changes to pastoral lease conditions, with partial success. Pers. comm.
38. Department of Primary Industry and Regional Development (2017) *Rangelands of Western Australia.* WA Department of Agriculture and Food, Perth.
39. Ratcliffe FN (1953) *Flying Fox and Drifting Sand.* Angus and Robertson, Sydney.
40. Holmes JH (1983) Extensive grazing in Australia's dry interior. In *Man and the Environment: Regional Perspectives.* (Eds RP Simpson, JH Holmes). Longman Cheshire, London.
41. EMA, P4227. Photographer unknown.

42. Barr N (2011) *The House on the Hill: The Transformation of Australia's Farming Communities.* Halstead Press, Canberra.
43. Van Etten EJ (2013) *Changes to Land Tenure and Pastoral Lease Ownership in Western Australia's Central Rangelands: Implications for Co-operative, Landscape-scale Management.* Edith Cowan University Publications, Perth.
44. Richards JD, Short J (1996) History of the disappearance of native fauna from the Nullarbor plain through the eyes of long-term resident Amy Crocker. *Western Australian Naturalist (Perth)* **21**(2), 89–96.
45. Hannet E (2007) Retired farmer. Interviewed by Nicole Chalmer, Esperance, 19 January.
46. Short J, Calaby JH (2001) The status of Australian mammals in 1922: collections and field notes of museum collector Charles Hoy. *Australian Zoologist* **31**(4), 533–562. doi:10.7882/AZ.2001.002
47. Johnson D, Johnson D (2014) Retired farm managers. Interviewed by Nicole Chalmer, Esperance.
48. Flannery T (2003) *Beautiful Lies: Population and Environment in Australia.* Black Inc., Melbourne.
49. Vega E, Montana C (2011) Effects of overgrazing and rainfall variability on the dynamics of semiarid banded vegetation patterns: a simulation study with cellular automata. *Journal of Arid Environments* **75**(1), 70–77. doi:10.1016/j.jaridenv.2010.08.001
50. Burbidge A, McKenzie NL (1989) Patterns in the modern decline of Western Australia's vertebrate fauna: causes and conservation implications. *Biological Conservation* **50**(1–4), 143–198. doi:10.1016/0006-3207(89)90009-8
51. Chalmer NY (n.d.) Cattle grazing in typha swamps on Esperance cattle properties. Pers. obs.
52. Pickering R (2013) *Australasian Bittern in Southwest Australia.* Department of Environment and Conservation, Perth, and Birdlife Australia, Melbourne.
53. Lawes MJ, Fisher DO, Johnson CN, Blomberg SP, Anke SK, Frank SA *et al.* (2015) Correlates of recent declines of rodents in northern and southern Australia: habitat structure is critical. *PLoS ONE* **10**(6), e0130626.
54. Fisher DO, Blomberg SP, Owens IPF (2003) Extrinsic versus intrinsic factors in the decline and extinction of Australian marsupials. *Proceedings of the Royal Society of London. Series B, Biological Sciences* **270**(1526), 1801–1808. doi:10.1098/rspb.2003.2447
55. Johnson CN, Isaac JL, Fisher DO (2007) Rarity of a top predator triggers continent-wide collapse of mammal prey: dingoes and marsupials in Australia. *Proceedings Biological Sciences* **274**(1608), 341–346.
56. Dunlap TR (1999) *Nature and the English Diaspora: Environment and History in the United States, Canada, Australia and New Zealand.* Cambridge University Press, Cambridge.
57. *Dog Act 1883* and *Dog Act 1903*.
58. Stephens D (2011) The molecular ecology of Australian wild dogs: hybridisation, gene flow and genetic structure at multiple geographic scales. PhD thesis, University of Western Australia, Perth.
59. Sharp A, Gibson L, Norton M, Ryan B, Marks A, Semeraro L (2002) The breeding season diet of wedge-tailed eagles (*Aquila audax*) in western New South Wales and the influence of rabbit calicivirus disease. *Wildlife Research* **29**(2), 175–184. doi:10.1071/WR00077
60. Frost W (1998) European farming, Australian pests: agricultural settlement and environmental disruption in Australia, 1800–1920. *Environment and History* **4**(2), 129–143. doi:10.3197/096734098779555682
61. Probandus (1878) Our rats and pordies. Letter Bag to the Editor. *Eastern Districts Chronicle*, 20 April, p. 3.
62. Sliprail B (1915) The Game Act and royalties. *Western Mail*, 5 February, p. 5.
63. Diamond J (2005) *Collapse: How Societies Choose to Fail or Succeed.* Viking Press, New York.
64. Abbott I (2006) Mammalian faunal collapse in Western Australia, 1875–1925: the hypothesised role of epizootic disease and a conceptual model of its origin, introduction, transmission, and spread. *Australian Zoologist* **33**(4), 530–561. doi:10.7882/AZ.2006.024
65. Martin G (2001) The role of small native mammals in soil building and water balance. *Stipa Native Grasses Newsletter* 16, 4–7.
66. Noble JC, Gillen J, Jacobson G, Low WA, Miller C, Mutitjulu Community (2001) The potential for degradation of landscape function and cultural values following the extinction of mitika (*Bettongia lesueur*) in central Australia. In *Land Degradation: Papers Selected from Contributions to the 6th Meeting of the International Geographical Union's Commission on Land Degradation and Desertification*, Perth, 20–28 September 1999. (Ed. A Conacher). Kluwer Academic, Dordrecht.

8
Civilising the bush

> Whilst I stretched my weary length along under the pleasant shade I saw in fancy busy crowds throng the scenes I was then amongst. I pictured to myself the bleating sheep and lowing herds wandering over these fertile hills; and I chose the very spot on which my house should stand ... I knew that within four or five years civilisation would have followed my tracks, and that rude nature and the savage would no longer reign supreme over so fine a territory.[1]

Explorer George Grey's optimistic predictions romanticised the Anglicised world and economy that he and other colonial explorers and settlers believed would progressively replace the existing landscapes of nature and Aboriginal social ecological systems (SES). His predictions came true – within 15 years the pastoralist vanguard had invaded. But in various regions throughout Australia extensive pastoralism was quickly replaced with closer settlements that promised new forms of economic prosperity for European settlers. In the Esperance bioregion, Aboriginal SES were mostly diminished by the period starting in the 1890s. Eventually, people of Ngadju and Nyungar descent worked for farmers, clearing the land and aiding the development of Anglo farming systems.[2,3]

In analysing impacts on Australian lands, Ian Reeve has identified three types of agriculture that featured in Australia since colonisation:

- pastoralism – featuring a high level of food self-sufficiency on the pastoral stations where fruit, vegetables, eggs, milk and meat were produced and often supplemented with wild animal food;
- closer settlement schemes – reflecting the yeoman ideal and significant local food self-sufficiency, continuing to about the 1930s;
- highly mechanised large-scale industrialised farming systems – developed after World War II, where regional food self-sufficiency is lost and monetary income is needed buy food, clothing, consumables and external farm inputs.[4]

There was a fundamental cultural desire and drive to reshape land into Anglo/European concepts of agricultural landscapes managed for food production and money. Reshaping was also driven by socioeconomic factors outside farmer control, such as international commodity prices and government political and economic agendas.[5] As transformations continued, the rapid downward trend in functional ecosystem biodiversity started by pastoralists and their overgrazing with sheep and cattle, continued with wide-scale clearing for farming.

Pastoral to freehold: dividing and renaming the landscape

Australian pastoralists effectively colonised and familiarised landscapes with geographic acts of naming, defining station boundaries and travel routes for people and their domestic animals.[6,i] By 1891, with John Forrest as Premier of Western Australia, the new aim became intensification and greater income from farming rather than pastoralism. Pastoral regions were surveyed into defined properties for intensified agricultural land use, with crops and livestock and closer settlement in areas perceived as reliable rainfall zones, to as low as 250 mm. Australia-wide inclusion of these low rainfall zones was uncondemned because many people shared a belief articulated by Nebraskan land speculator Charles Dana Wilber who, in encouraging the development of the Great Plains in the US, declared that 'rain follows the plough'.[7]

In the municipality of Esperance, the closer settlement zone radiated out around the town for approximately 180 km. The district started east of Munglinup (a town 180 km west of Esperance) and ranged to Israelite Bay, from Cascades north across to Mardabilla Rock, within the Eucla Land Division, one of the five Land Divisions in Western Australia. Salmon Gums, at this time and until 1989, was included in the Land Division of Dundas.[8] By the mid-1890s smaller farms were developed in the fertile patches of the coastal sandplain. Farmers were using intensive methods to produce higher-value food, including dairy, eggs, honey and poultry as well as horticultural produce for local consumers. In the Esperance mallee bioregion, farms were established as an alternative to pastoralism by 1905. This had also occurred in South Australia, Victoria and NSW mallee lands, where smaller-scale properties were developed and cleared of native vegetation to grow wheat crops for export. There was a concerted political and social mandate to do this in Western Australia and before 1949 most larger-scale land subdivision and landscape re-engineering had taken place in Esperance mallee lands.[9]

Post-settlement, government surveyors throughout Australia began developing methods to assess the suitability of land for cropping, and gave it a rating determined by soil appearance and the types of native vegetation growing upon it. In Western Australia, these surveyors' criteria were developed by the Lands and Surveys Department to apply to big land releases in the south-west of the state in the early 20th century.[9] 'First class' land was considered analogous to the red brown earths of eastern Australia, but in Western Australia they included the degrading granite soils, vegetated with eucalypt and acacia woodlands. 'Second class' land featured soils with grey to brown sand and sandy loam topsoils with tough sandy clay subsoils. The rounded limestone nodules in these soils are an example of bio-pedogenesis (plants making soils) by trees of different species. They are found in lower rainfall areas where native vegetation consists of eucalypts and mallee eucalypts of various species.[10] 'Third class' land consisted of the extremely infertile sandy and ironstone gravel soils growing hard-leafed shrubs 1–3 m high (banksia heath, kwongan and mallee heath of the sandplain) with clay increasing at depth and in some places a distinct clay horizon. There was no equivalent class in the south-eastern states. Another category with no equivalent was 'Poison land' infested with *Gastrolobium* species; it was coloured black on the subdivision maps.[11]

---

i Gill asserts that the opening up acts were predicated on seeing the land as empty and available.

In 1896 surveyor E.S. Brockman reported to the Crown Lands Department about a survey he had completed to ascertain the agricultural potential of the Esperance region. He accurately described the country 50 miles (80 km) on each side of Esperance as:

> ... firstly a five mile [8 km] wide strip of sand and limestone hills grassed with coarse coastal grasses and numerous little hollows with good soils and water suitable for vegetables, root crops and lucerne; adjoining this strip to the north is an irregular belt about 25 miles [40 km] wide of sandplain country of loose sand and gravel, poorly watered with poor vegetation though 'certain rainfall'; beyond in lower rainfall is mallee scrub growing on poorer sandy gravelly soils which gradually improve until 60 miles [96 km] further from the coast to Norseman are generally good soils with mallee, salmon gums and black-hearted forest and well grassed in patches. In general, the country is poor where rainfall is good and good where rainfall is poor.[12,13]

Brockman noted that in the sandplain there were very good soils in the river valleys where small rivers that arose in the mallee passed through the sandplain and coastal dunes to the sea. These strips were found along the Oldfield (west of Munglinup), Munglinup, Young, Lort and Thomas Rivers and in patches along the mallee–sandplain boundary (bio-pedogenesis occurred there and along the river valleys). He was very impressed with the well grassed, lightly treed Dalyup River valley in a good rainfall zone. The areas of Dalyup along the Dalyup River, Myrup on Bandy Creek and Doombup on Stockyard Creek were chosen for intensive subdivisions because of their 'First class' soils.[12]

In Brockman's opinion, the size of a property was to be determined by the amount of land necessary for a European settler and family to make a good living using the agricultural systems the surveyor decided was most appropriate to the dominant land class.[12] For example, Dalyup River's 'First class' land, with its clays and strong loams, merited small subdivisions of 100 acres (40 ha) for high-return, intensive mixed horticultural agriculture.[14] In the past, those quality lands had been the focus of Aboriginal SES and were the Dempster brothers' first choice in 1864 for their Mainbenup homestead. In 1868, however, the brothers relocated to Dempster Head (now Esperance town) for better shipping access.

Fencing property boundaries was one of the first jobs for farm settlers everywhere in Australia. Fences were a visible and concrete manifestation of the surveyor's plan. The men who surveyed huge landscapes, usually into a grid pattern of rectangular blocks, rarely acknowledged landscape features or impediments. Their reality was expressed as two-dimensional lines on paper. Situated about 20 km east of Esperance, on the very large granite monolith Mt Merivale, a title survey line assigns two-thirds of the monolith to one title and one-third to another title – fencing over it is unworkable. At my farm at Coronet Hill, an original title survey line can't be fenced as it goes through a 100 ha lake![15,16]

Agriculture, grass and the yeoman ideal

In England, 'yeoman' farms based on mixed animal husbandry, grass and crops, fruits and vegetables and green manuring were long enduring and sustainable SES. Comparable to

Australia, the English landscape was a complex web of interrelationships between families, farms, animals, nature and their SES. The common lands were managed with strict rules and customs to prevent abuse, mismanagement and cheating.[17] In 1873 British nature writer Richard Jefferies summarised how a typical self-sufficient yeoman farm functioned:

> The typical farmhouse … contained … half a dozen trades – bread baked from the farm's own wheat, meat from its own animals, slaughtered, butchered and cured on the farm; carefully managed areas of farm woodland, keeping the farm in fuel, tools, fencing materials and building timber; manure supplied from the debris of old crops or from livestock; the very rotation of crops so arranged as to preclude any exhaustion of the soil …[18]

The European invasion and settlement of Concord, New England in the United States was modelled on British yeoman principles. Environmental historian Brian Donahue explains how the soil nutrients for the yeoman system were supplied by grass growing on the fertile river flat soils, continually replenished by silt and mineral nutrients from the mountain catchments. It was as essential in the yeoman SES as it had been to the Native American SES they displaced. Livestock grazed the fertile grasslands, which also provided the hay that fed the livestock shedded over winter. The animals provided the manure used in spring to fertilise crops. Keeping the balance between arable (grain crops) and grassland was critical, because livestock were indispensable in transferring mountain nutrients to arable land via manure, for a sustainable system.[19]

Such systems had lasted for centuries in England. They only failed when the balance between grain production and pasture, livestock and manure was broken as grain production land increased at the expense of the other components. This was to feed erupting human populations as industrialisation expanded and grain became highly profitable. The sustainable agroecological system in New England also failed when industrialisation took hold and the arable–pasture balance was lost.[19] This pattern is being repeated around the world. Industrialisation and erupting human populations can no longer rely on sustainable closed nutrient-loop agroecological systems. Instead, they produce for export and local use by buying in artificial fertilisers. This industrialised type of production only became possible when fertilisers (guano was the first used) were available to bring in from elsewhere.

In Australian colonial agriculture, a natural nutrient recycling system was difficult for soils were infertile and there were few mountain catchments. Early Australian cropping agriculture jumped straight into using artificial fertilisers, though livestock usually grazed native pastures. D.C. Markey's research concerning foodways in colonial Western Australia, concludes that much of the push for new land was due to degradation and exhaustion of soil fertility in lands settled early by Europeans.[20] In 1962, Donald Meinig's research in South Australia stressed that early European farmers focused on producing agricultural commodities (wool and wheat), selling to distant places and removing nutrients from already poor soils. They used the money to buy life essentials such as food from elsewhere. He concluded that early European farmers in Australia did not come from a farming background, viewed the land as hostile and had little knowledge of how to farm without running the system down as they exhausted the soil.[21]

Biophysically sustainable yeoman farm systems that recycle nutrients need the younger fertile soils of Eurasia and North America. Younger soils derived from geologically recent glaciations and vulcanism hold large amounts of macro- and micronutrients to maintain the level of nutrient export that happens when produce is transferred off farm. The paucity of recent geological soil renewal means that most Australian soils are comparatively infertile, particularly in south-western Australia.[22] Without artificial manuring, there was no capacity for a sustainable yeoman system in Western Australia, as the Final Report of the Commission on Agriculture concluded in 1891.[23]

In the Esperance bioregion, the mobile pastoral system could profitably produce wool. Fruit and vegetable crops for home consumption used animal manures but there were many difficulties doing this on a larger scale, especially as stock were not confined for lengthy periods.[24] Crops of wheat were grown on about a third of the Campbell Taylor enterprise at Thomas River Station, mainly as feed for horses.[25,26] This could only be done on the intrusions of better-quality mallee-type soils created by bio-pedogenesis and on the river flats with alluvial soil that were a feature on this station. However, the nutrient base would have gradually eroded unless replaced with phosphorus-based fertilisers, potassium and nitrogen. Herbivore manures are poor in phosphorus and, unless an animal dies on-site and is allowed to decompose into the soils where it fed, the phosphorus it contains is permanently removed when it is sold. Selling 1000 kg of beef removes about 7.2 kg of phosphorus from the land the beef cattle grew upon.[27,28] This livestock nutrient removal may be an unacknowledged contributor to rangeland degradation in Australia.

In Britain, as population growth and industrialisation continued, food self-sufficiency was lost and food imports became essential. By the early 20th century 80% of the nation's wheat, 60% of its fruit and 40% of its meat was imported, much from Australia and New Zealand. Economic returns from these export commodities furthered the desire to clear and farm more land in the colonies.[23,29] Despite the disappearing yeoman system in Britain and the poor local soils, this form of land use and the social systems it embodied inspired Western Australia's agricultural development policy in the years following the granting of self-government in 1891. John Forrest, Premier for 10 years until 1901, drove the yeoman ideal, introducing various legislative means to ensure its progress.[30]

Forrest placed a high priority on producing wheat for local self-sufficiency in grain and flour, as both were being imported from South Australia.[31] In 1887 H.W. Venn chaired a three-year royal commission into agriculture. Its findings supported the great agricultural push of the early 20th century that continued beyond Forrest's time in office. It advocated that Western Australia was suitable for wheat-growing and dairying and for the planting of softwoods, that it needed an agricultural bank and agricultural education, and that the government should repurchase and subdivide large pastoral estates.[32,33] In response the government passed legislation, based on US homestead legislation, called the *Land Regulations and Homestead Act 1893*.[34] Any person who resided on the land, cleared it and met a minimum target of improvements each year would be entitled to apply for a free grant of 160 acres (almost 65 ha).

George Throssell, Minister for Lands, asserted in 1897 that '160 acres and a wife were all that was required to make the majority of Perth young men happy'. The Act also made changes to Conditional Purchase regulations, with details varying between regions. More remote regions could apply for larger land holdings and more generous terms than those in the south-west.[35] With no agricultural lending by private banks, a state-owned Agricultural Bank was formed with the purpose of allowing cash-poor farmers to borrow money on favourable terms.[5]

Developing the sandplains

Land surveys subdivided much of the original pastoral holdings in the Esperance bioregion. Smaller, more intensive farms producing food for local and Goldfields region consumption were established. These first farms were in the sandplain, which had plenty of rain and fresh water. Soil nutrient deficiencies restricted farms to fertile areas such as silt-accumulating swamps, degrading granitic soils around granite outcrops, red earths on river and creek flats and patches of mallee-type soils.[36]

The development of the mining industry in the Goldfields region in the late 1890s provided a market for intensively farmed agricultural produce. The number of horses used for mining transport led to enormous demand for oats, hay and chaff, which comprised the major income source for early farms.[37] Intensive farms with horticulture, pigs, poultry and dairying developed along the well-watered quality soils of Dalyup River, Bandy Creek and Doombup Creek, fulfilling the state government's yeoman farming aims to produce for local towns – in this case for Esperance and Goldfield townships – albeit with applications of bought-in fertiliser.[5,8]

The farm established by the Stewart family was an example of this new type of food producing system. In 1896 the first dairy was established in Esperance by Sarah Stewart (the driving force) and her husband Robert. An excerpt from the *Esperance Times* reported that milk was available for 4 pence per pint (475 mL).[8,38] By 1897, the pair owned or leased around 2000 acres of land at Dalyup which they named Park Farm. It had fertile riverine soils and less fertile sandplain country. Bush was cleared and orchard and vegetable gardens established with water supplied by the first underground well in the district. By 1898 Park Farm was a mixed farm, virtually self-sufficient with food and nearly all everyday items produced there.[39] Crops included wheat, maize, oats, potatoes and a range of vegetables. The farm used some animal manures but it was the first in the district to import and use guano and phosphate fertiliser.[39] The orchard provided quantities of fruit; there was farm-grown meat from cattle, sheep and pigs; and ocean-caught fish were salted, smoked and pickled. The dairy produced milk, butter and cheese, and salt was gathered from nearby salt lakes.[40,41] Sarah drove a large buggy into Norseman to sell excess fruit and vegetables – a round trip of about 400 km.

Early farmers in the Merivale district east of Esperance had grazing leaseholds scattered around the district. They continued a form of pastoral mobile grazing as stock was moved between pastures on good soil patches, that helped in preventing livestock coastal disease. European settlers supplemented their food and income by eating wildlife and selling products such as kangaroo skins. Bush plant foods were also eaten but to a lesser extent, though quandongs seemed to be eaten everywhere they grew in Australia. Clothes were homemade and wagga bed rugs were made with wallaby and possum skins.[36]

In relation to early settlement in the western US, researcher Michael Merrill concludes that 'farmers in isolated mountain areas or in frontier regions during the first year or two were self-sufficient', growing and/or hunting all their food and making their own clothes and implements. Necessities intimately connected to civilisation, including flour, sugar, tea and cloth, reflective of social status and culture, had to be bought as part of the cash economy, providing a powerful incentive to earn money.[42] Merrill found that this 'orientation to the market was the most important characteristic of early American farmers'. Early European farmers throughout Australia were similarly motivated to become part of the state, national and international economies.[43]

When visiting Esperance, the 1887 Commission into Agriculture found that farms were generally self-sufficient in food but cash markets for local produce were hard to come by.[44] As in modern times, local produce was competing with far cheaper imports from overseas and interstate. Storekeeper William Hughes explained to the Commission that though he bought as much local produce as possible, it was more expensive to cart local produce from the mallee than to import it from Albany by ship. Despite this, the statistical returns from 1896 show that expanding varieties of produce were being grown for local Esperance town consumption.[45] Farmers explained that using bone dust and guano (both contain concentrated phosphorus) improved soil fertility and yields dramatically.

No phosphorus, no farming
The 1891 report produced by the Commission on Agriculture (which sat from 1887 to 1891) described farming in many districts of south-western Australia on different classes of land. It emphasised extensive stretches of sandplain and ironstone country as 'useless country, unfit for almost anything, intersecting the good land'.[44] With such early recognition of the infertility of Western Australian soils, it became accepted that fertilisers containing phosphorus were essential to grow plants and animals for European-type agricultural systems.[44] As Sarah Stewart related to a later Royal Commission, in the early days crops could only be grown for one year without the addition of bone dust or phosphate fertiliser.[40] This infertility was also apparent in other states – trials conducted at Roseworthy College in South Australia in 1882 showed that wheat made a remarkable response to superphosphate. Progressive farmers in South Australia began importing phosphate fertilisers before local production of superphosphate started.[46]

In Western Australia, the first commercial fertilisers were based upon guano deposits mined on the Abrolhos Islands. In 1900, phosphatic fertilisers were imported from England to Western Australia and phosphate had been discovered on islands of the Recherche Archipelago, as outlined by the 1908 Geological Survey Report.[47] A proposed venture to mine phosphate stalled when higher-quality and cheaper phosphate was sourced from Christmas Island south of Java, in the Indian Ocean. The island has been extensively mined for phosphate since 1897, when the Christmas Island Phosphate Co. Ltd was formed.[48] Today there are no commercially viable amounts left on Christmas Island. Australia must now source phosphate from other countries.

A 1908 report by the Agricultural Department of Western Australia describes experiments to bring the extensive sandplains into agriculture, 'it is ascertained by analysis,

[they] are very deficient in all plant foods, but especially in nitrogen and phosphoric acid ... The lack of humus and low capacity for moisture of these lands rendered it necessary to effect radical alterations in their physical character'.[49] Early resistance to large-scale sandplain land-clearing was due to fears of wind erosion of the light sandy soils. This concern helped decide the government decide that the state's first large-scale cropping land releases should comprise heavy soil mallee country between Grass Patch and Salmon Gums.[8]

The 1916 Royal Commission on the Agricultural Industries of Western Australia prepared a map defining the area considered to be the safe rainfall limit for growing wheat.[49] This line became known as the Brockman Line after Surveyor-General F.S. Brockman; however, as it was not based on long-term rainfall data it failed to do more than define current wheat-growing areas. It appears to have had little impact on later mallee developments.[50]

Farming the mallee woodlands

The woods and their trees were an important part of the early English agrarian world. Trees provided fuel and materials for many products ranging from houses to stockyards, bridges to ships. Despite their importance, little effort was made to sustainably harvest and reforest and by 1350 clearing for agricultural expansion (largely financed by sale of timber) had left only 10% of England wooded.[51] This reflected a gradually shifting baseline where the long, slow disappearance of woodlands did not activate tree replanting or conservation. Unlike in Australia, clearing did not quickly result in obvious environmental problems, such as salinisation. This ethos of clearing for profit and agriculture was therefore deeply embedded in the colonising culture and accepted as a normal process of landscape 'improvement'. There was also a common-sense belief that the best agricultural land was reflected in the size and density of trees growing in a landscape. Early European settler experience in the Esperance bioregion sandplain had proved this correct – treed soils of yate swamps, river and creek systems were the most fertile.[41,52,ii] The mallee region has more fertile soils than the sandplain. However, this belief does not always apply to high rainfall areas. Denmark and Walpole on the western south coast have over 1200 mm of rain annually but the huge trees – karri (*Eucalyptus diversicolor*) and the three tingle species (*E. jacksonii, E. guilfoylei, E. brevistylis*) – grow in extremely nutrient-deficient soils. As in other large tree temperate areas, tropics and sub-tropics in Australia and worldwide, nutrients are tied up in the vegetation.[53]

The first large-scale farm in Western Australia's mallee region was Grass Patch Farm 79 km north of Esperance, so named for its large tracts of open grasslands – likely a result of previous Aboriginal management. It comprised 3783 acres (1530 ha) developed from 1894 by the Esperance Proprietary Co., after the 1892–1893 gold discovery at Norseman, Coolgardie and Kalgoorlie. The aim was to supply horse-drawn transport of people and supplies, and hay and oats for Goldfields horses. Oats were first sown into the grasslands in 1895 and later, horses were bred to supply the transport company Cobb and Co.[54] Cobb and Co. and Gilmores provided the main transport, with coaches and goods wagons, for the large stream of traffic from Esperance Port to the Goldfield diggings. Their horse teams were

---

ii   West has found that islands of fertility occur around trees and shrubs in arid landscapes.

fed, rested or changed at regular staging posts along the route at 25–50 km intervals. With no fresh water available, water from salt lakes in a landscape of trees was distilled from wood-fired condensers at the staging posts.[55]

Grass Patch Farm was the ideal yeoman farm, with a mix of farm enterprises producing crops and livestock for income and self-sufficiency. It was successful beyond the government yeoman vision because it was owned by people sufficiently capitalised for development and buffered for hard times, its soils were relatively fertile and rainfall was relatively reliable. Being much larger than the meagre 100 acres considered sufficient by government, it could afford to employ workers, and from establishment it had a nearby cash market.

Population growth and loss of state revenue from gold after 1910 helped drive the second pioneering era in Western Australia that aimed for export income and self-sufficiency in wheat.[56] There was a populist view that Australia was a vast empty land with unlimited potential that could be filled with enormous numbers of people, perhaps 100–500 million.[57] These predictions were so firmly held that when disputed by University of Sydney geographer Griffith Taylor, a geography textbook he wrote about Australia was banned by the Western Australian Education Department in 1921 because it emphasised the predominant influence of aridity. Taylor eventually left Australia for the University of Chicago because of the national hostility to his realistic science-based views on the limits to growth. Such modern science-based views are often still dismissed.[57]

This over-optimistic form of thinking took further hold when, supporting a 1911 Parliamentary Bill to construct a railway from Norseman to Esperance, the then Attorney General Thomas Walker extolled the suitability of the mallee region north of Esperance for smaller-scale intensive settlement:

> … we have some of the richest land upon the surface of our globe; and given the application of scientific farming, on dry farming we could support populations as big as those that flourished once in the valley of the Euphrates and built Babylon.[58]

Such political 'boosterism', where subjective beliefs override scientific reality, has often been a feature of the under-informed and environmentally destructive types of development imposed upon landscapes throughout Australia, which leave future generations to solve the resultant problems.[iii]

Learning a complete lesson from history was overlooked in Walker's pronouncement about Babylon – which eventually disappeared, as did its population. Research has confirmed that Babylon failed due to environmental collapse from desertification and salinity caused by unsustainable attempts to grow more grain to feed the erupting population.[59] Grain growing has become the most important agricultural enterprise in Western Australia and in significant areas of eastern Australia, not so much to feed the local millions as to feed the millions erupting overseas. And like at Babylon, salinity, creeping desertification and eventual fertility challenges are major environmental problems.[60]

---

iii Australian boosterism refers to potential development claims unsupported by real research – to the extent that the idea is believed to reflect reality and dissenters are castigated.

Following the success of Grass Patch Farm, more land north and north-east of Esperance in Grass Patch, Red Lake, Scadden and Circle Valley was surveyed into farms and settled. Financial help from the state Agricultural Bank was given for clearing, dam building, fencing, seed wheat and fertiliser for the first crop. Unfortunately, cereal yields were very poor and many European settlers walked off their farms as low yields due to intrinsically saline soils were compounded by poor prices and the 1914–1915 drought.[61] In response to this failure, a Royal Commission was held during September 1916, to find out if successful farming in the state mallee belt and Esperance lands was possible. A positive finding would provide evidence to support construction of a railway from Kalgoorlie to Esperance. There were numerous farmer interviews and expert opinions, including that of Government Chemical Analyst E.A. Mann, who had found that 96 of the 128 Esperance mallee soil samples he analysed contained over 0.05% salt.[62] Further testing by J.W. Patterson, Professor of Agriculture at the University of Western Australia, confirmed Mann's results:

> The results of the analyses ... indicate that many of these soils ... contain too much salt to make wheat farming profitable ... About one third of the agricultural area in the district does not contain too much salt for settlement; about one-sixth is doubtful, and about one-half of the area contains too much salt for profitable farming.[63]

The Commission preferred to ignore both reports and instead followed the positive findings for Victoria made by the Pinnaroo Royal Commission of 1905.[64] That report accepted that improvement of Victorian mallee lands occurs with age, taking at least six to seven years, and yearly use of superphosphate. Following those findings, the development of the Esperance mallee to grow wheat was approved. Extensive settlement post World War 1 by ex-servicemen, civilians and migrants was encouraged on farms of 1100 acres (440 ha).[65] After visiting the region in 1923, the Minister for Agriculture, H.K. Maley, made unsubstantiated statements that wheat could be grown in arid Higginsville, almost 130 km north of Norseman (mean rainfall 289.5 mm), once the projected rail line was connected.[66]

So it started, and the mallee underwent relentless clearing. Horse-drawn rollers and steam-driven tractors dragging even larger rollers, were used (see Fig. 8.1). So many horses were needed for this scale of landscape re-engineering that large numbers were imported from eastern Australia. Farmer Albert Kent from Dalyup described how horses were used well into the 1940s, as tractors were scarce and expensive.[67] Karl Dimer described how many of the horses escaped and were rescued by his family at Nanambinia Station as they attempted to trek the thousands of kilometres eastwards to home, across the Nullarbor.[24]

From 1919 an extensive mixed farming zone was established along the Esperance–Kalgoorlie rail line, that was completed to Norseman by 1927 and linked the region to the rest of the state. Esperance Port exported grain and produce and imported superphosphate.[8]

State policy was at a crossroads, as the effects of land clearing on salinisation of water bodies and soils were unmistakable.[69] There were two options – believe the science or follow the unsubstantiated pronouncements of non-scientists, such as those of opposition leader Sir James Mitchell who vigorously opposed the use of soil science to decide upon land release

Fig. 8.1: Clem Holt clearing the mallee using horses and a boiler from the goldmines. Location 1027, East Dowak, Salmon Gums.[68]

areas.[70] Though Salmon Gums Research Station had been established in 1926 to conduct research on crop growing in existing and future land developments, in 1927 the new state government led by Mitchell was too impatient to await its research findings. Mitchell advanced his non-evidence-based agenda, successfully proposing that all lands between Southern Cross and Salmon Gums, comprising about 8 000 000 acres (3 237 485 ha) of Western Australia, should be cleared for agriculture.[71] With expectations that this would create 3500 new wheat and sheep farms, the project became known as the 3500 Farms Scheme.[72] The state government proposed it as a soldier settlement scheme to be included in the British and Commonwealth governments' migration policy plan to bring more immigrants to Australia. This would relieve the ongoing British population explosion and populate the colonies with 'true Englishmen'.

The scheme was opposed by the state's Agricultural Bank. It had seen first-hand the numerous production problems of the regions' existing farmers, who were unable repay loans. L.J.H. Teakle, Plant Nutrition Officer for the state Department of Agriculture, was called in to review the land quality. His report concluded that the soils were either too saline or alkaline for optimum wheat growth. This prompted the withdrawal by the British government in 1932, and the 3500 Farms Scheme was then abandoned.[72] Further worsening the issues, a drought had started in 1928, causing plummeting yields and extreme water shortages that stressed livestock and households. In the strange synchronies that often plague farming, prices also collapsed from 1929–1930, lasting many years as the Great Depression deepened.[8]

This awful period of environmental, personal, social and economic stress led to the abandonment of over 75% of farms in the Esperance mallee. These shocks show up in state records with the area under wheat cultivation declining significantly from 40 525 acres in the 1929–1930 season to around 6310 acres by the 1936–1937 season.[73,74] With the abandonment of farms and their ownership reverting to the Agricultural Bank, wild nature briefly returned. Horses left behind in the exodus went wild and the bushlands regrew. Some people were environmentally aware enough to live with the reality of mallee lands – the Griggs were a family who remained and made a living as caretakers of the abandoned properties. They resumed the sustainable practice of mobile pastoralism, rotating cattle through the southern Scadden blocks and burning grassy areas periodically to prevent tree regrowth and provide fresh feed for the cattle.[72]

The farm abandonments became an opportunity for future farmers to acquire large acreages at the cost of the debt to the Agricultural Bank. Albert Kent described how as young man he bought his 4000 acre (1620 ha) farm at Circle Valley in 1952 by paying off the debt on the four blocks that comprise it.[67]

Salty soils

In Western Australia and throughout Australia, primary salinity has developed naturally over a long period of time, mainly in areas where rainfall is insufficient to leach salts from the soil profile and evaporation is high. This form of salinity is a feature of the heavy textured, highly alkaline and usually well drained soils of the Australia's mallee lands. High subsoil salinity, in contrast, originates from sea salt deposited during the Eocene epoch when the land was covered by a shallow sea, as well as salt from thousands of years of rainfall. In Western Australia (bio-pedogenic research seems lacking in other states) surface soil salt is also concentrated in a bio-pedogenesis process by eucalypt species such as red morrel (*Eucalyptus longicornis*) and black morrel (*E. melanoxylon*). They concentrate salt around their roots to maintain a competitive advantage.[75,76,77] These naturally salty soils are prime candidates for secondary salinity as water tables rise due to the clearing of permanent perennial vegetation. This destroys the hydrological balance.

During his investigations for the 3500 Farm Scheme, Teakle had found that sandy soils were the lowest in salt and that poor crops and pasture growth on many mallee soil types were due to high salt levels in the top 15 cm of soil.[61] Despite these findings land development continued and surprisingly, as happened in the Victorian mallee, in some soils salt levels gradually decreased after clearing. This is hypothesised to be linked to greater runoff and percolation of rainwater after the trees are removed. It may also reflect the cessation of salt-forming bio-pedogenesis by salt-excreting trees and other vegetation.[75]

The program of soil surveys and mapping continued until 1938, with Teakle spending vast amounts of time in the mallee woodlands. His observations that the ecosystems were surprisingly poor in wildlife may reflect past habitat destruction impacts of pastoralism, ending of management by Aboriginal SES and an influx of foxes at this time. The soil surveys resulted in the first comprehensive regional classification of soils in Western Australia. The document contained standardised soil profiles, examined relationships between morphology and chemistry and defined the vegetation associations for different soils. Teakle's suggestion that some eucalypt species may have a role in lime accumulation in the soil surface where they grew, was an early notion – since proven correct – of the bio-pedogenesis concept.[61]

A reconstruction plan for mallee farming regions based on Teakle's work was developed in the 1940s. Land surveys were realigned to make farms larger, with a minimum of 2000 acres (809 ha), farmers were to avoid clearing salty soils, each farm had to have enough favourable soils to be viable. Farm systems shifted from wheat only to mixed farming (cereals, lupins and sheep) that were intended to make better use of soils not suited to wheat, and a measure of economic (and ecological) resilience by diversifying enterprises.[8]

The soil science research of this period in Western Australia should have been recognised as a significant turning point in how to decide which land could be developed for farming, and how. The lack of vision about alternative farming systems that kept perennials in landscapes set a narrow-minded trajectory that still dominates. Pre-emptively completing land capability assessments based on soil research could have reduced both environmental damage and socio-economic damage to farmers. Unfortunately, the attitude of 'deal with consequences later' prevailed, and plans were made for the environmental engineering and massive transformation of the soils and sandplains of Western Australia. It would be the feature of industrialised agricultural expansion in Australia after World War II.

Endnotes

1. Grey G (1841) *Journals of Two Expeditions of Discovery in North-West and Western Australia, during the Years 1837, 1838 and 1839, Under the Authority of Her Majesty's Government, Describing many Newly Discovered Important and Fertile Districts, with Observations on the Moral and Physical Condition of the Aboriginal Inhabitants.* Vol. 2. T. and W. Boone, London.
2. Steer Y, Steer A (2016) Mother and son, retired farmers. Interviewed by Nicole Chalmer, Esperance, 7 November.
3. Dabb A (2018) Nyungar Elder. Interviewed by Nicole Chalmer, Esperance, 18 January.
4. Reeve IJ (n.d.) *A Squandered Land: 200 Years of Land Degradation in Australia.* Rural Development Centre.
5. Rintoul J (1986) *Esperance Yesterday and Today.* 4th edn. Shire of Esperance.
6. Gill N (2005) Life and death in Australian 'heartlands': pastoralism, ecology and rethinking the outback. *Journal of Rural Studies* **21**(1), 39–53. doi:10.1016/j.jrurstud.2004.08.005
7. Passioura J (2007) The drought environment: physical, biological and agricultural perspectives. *Journal of Experimental Botany* **58**(2), 113–117. doi:10.1093/jxb/erl212
8. Shire of Esperance (1989) *Esperance Yesterday and Today.* 4th edn. Shire of Esperance.
9. McArthur WM (1991) *Reference Soils of South-western Australia.* Department of Agriculture, Perth.
10. Galloway P (2012) Research Officer, Soil Science, Department of Agriculture and Food, Esperance. Interviewed by Nicole Chalmer, 7 December.
11. Henzell T (2007) *Australian Agriculture: Its History and Challenges.* CSIRO Publishing, Melbourne.
12. Angove WH (n.d.) Field Book No. 50 – AU WA A60 Crown Lands and Surveys Department (1873–01–01–1890–01–01) Series S237. State Records Office, Perth.
13. Lindley-Cowan L (1897) *The West Australian Settler's Guide and Farmer's Handbook.* E.S. Wigg and Son, Perth.
14. Meharry WT (1937) Land classification and subdivision. *The Australian Surveyor*, 1 December, 418–428.

15. Fahey C (2011) The free selector's landscape: moulding the Victorian farming districts, 1870–1915. *Studies in the History of Gardens & Designed Landscapes* **31**(2), 97–108. doi:10.1080/14601176.2011.556370
16. Chalmer NY (2016) Pers. obs.
17. Rebanks J (2016) *The Shepherd's Life: A Tale of the Lake District*. Penguin, London.
18. Porter R (1990) *English Society in the Eighteenth Century*. Penguin, London. 9. R Jefferies (1873) as quoted.
19. Donahue B (2004) *The Great Meadow: Farmers and the Land in Colonial Concord*. Yale University Press, New Haven.
20. Markey DC (1986) On the edge of empire: foodways in Western Australia, 1829–1979. PhD thesis, Pennsylvania State University.
21. Meinig DW (1962) *On the Margins of the Good Earth*. Rigby, Adelaide.
22. Smith DA (2000) *Natural Gain in the Grazing Lands of Southern Australia*. University of NSW Press, Sydney.
23. Venn HW (1891). *Final Report of the Commission on Agriculture*. Government Printer, Perth.
24. Dimer K (1989) *Elsewhere Fine*. South West Printing and Publishing, Bunbury.
25. Bridges J (2004) *Challenge in Isolation*. John Bridges, Esperance.
26. *Western Australian Yearbooks 1896–1905: Agriculture*. Government Statistician's Office, Government Printer, Perth.
27. Impact Fertilisers (2020) *Nutrient Removal by Crop: Animals*. <https://impactfertilisers.com.au/wp-content/uploads/impact-calc-nutrient-removal-chart.pdf>
28. Brown AD (2003) *Feed or Feedback: Agriculture, Population Dynamics and the State of the Planet*. International Books, Utrecht.
29. Ponting C (2007) *A New Green History of the World: The Environment and the Collapse of Great Civilizations*. Random House, New York.
30. Tonts M (2002) State policy and the yeoman ideal: agricultural development in Western Australia, 1890–1914. *Landscape Research* **27**(1), 103–115. doi:10.1080/01426390220110793
31. Henzell T (2007) *Australian Agriculture: Its History and Challenges*. CSIRO Publishing, Melbourne.
32. Bolton GC (2008) *Land of Vision and Mirage: A History of Western Australia since 1826* [online]. University of WA Press, Perth.
33. Bolton GC (1990) Venn, Henry Whittall (1844–1908). *Australian Dictionary of Biography*. Vol. 12. National Centre of Biography, Australian National University, Canberra.
34. Western Australian Government (1893) *Homesteads Act*.
35. Western Australian Department of Agriculture (n.d.) Land Laws. 3.
36. King C (2012) *The Making of Merivale*. Hesperian Press, Perth.
37. Glynn S (1975) *Government Policy and Agricultural Development: A Study of the Role of Government in the Development of the Western Australian Wheatbelt, 1900–1930*. University of WA Press, Perth.
38. *Esperance Times*, 12 October 1898.
39. Letter from Kath Rundell to Mrs Murray, 1976. EMA 1506, Battye Library, Perth.
40. K. Rundell (1979) *The Stewarts of Dalyup*. PR 9686/1. Battye Library, Perth.
41. *Stewart Family Notes*. n.d. PR 9686/1-, Battye Library, Perth.
42. Merrill M (1977) Cash is good to eat: self-sufficiency and exchange in the rural economy of the United States. *Radical History Review* **1977**(13), 42–71. doi:10.1215/1636545-1977-13-42
43. Muir C (2010) Feeding the world. In *Griffith Review 27: Food Chain*. (Ed. J Schultz). Text Publishing, Melbourne.
44. Commission on Agriculture 1891. Final Report of the Commission on Agriculture. In *Votes and Proceedings of the Western Australian Parliament*, Part XII. Government Printer, Perth.
45. *Western Australian Yearbooks 1886–1906: Agriculture*. Government Statistician's Office, Government Printer, Perth.
46. Loneragan JF (1999) Nutrients: a sparse resource. In *Plants in Action: Adaptation in Nature, Performance in Cultivation*. (Eds BJ Atwell, PE Kriedemann, CG Turnbull) pp. 300–345. Macmillan Education, Melbourne.
47. Woodward HP (n.d.) *Geological Survey of Western Australia 1908: Plan shewing Phosphate Deposits on Christmas Island [i.e. Daw Island] Recherche Archipelago Eastern Group*. Geological Survey of Western Australia, Perth.

48. *Christmas Island*. Fact Sheet 157. National Archives of Australia.
49. Royal Commission on the Agricultural Industries of Western Australia (1916).
50. Morgan R (2010) Settling seasons: climate and agrarian enterprise. *Early Days: Journal of the Royal Western Australian Historical Society* **13**(4), 514–533.
51. Bowles CR (1980) Ecological crisis in fourteenth century Europe. In *Historical Ecology: Essays on Environment and Social Change*. (Ed. LJ Bilsky) pp. 86–99. Kennikot Press, New York.
52. West NE (1991) Nutrient cycling in soils of semiarid and arid regions. In *Semiarid Lands and Deserts: Soil Resource and Reclamation*. (Ed. J Skujins) pp. 295–332. Taylor and Francis, London.
53. Nykvist N (1997) Do logs from tropical rain forests contain more plant nutrients than logs from temperate forests: a literature review. *Journal of Sustainable Forestry* **7**(1–2), 1–19. doi:10.1300/J091v07n01_01
54. Freeman K, Freeman B (1995) The grass patch. In *Faith, Hope and Reality: Esperance 1885–1995*. (Coord. P Blumann). Esperance Shire Council.
55. Campbell D (2006) Scadden. In *Pioneers and Early Settlers of Scadden, 1910–1959*. (Ed. L Bale). Publisher unknown.
56. Gill N, Anderson K (2005) Improvement in the inland: culture and nature in the Australian rangelands. *Australian Humanities Review* **34**. <https://ro.uow.edu.au/scipapers/4799>
57. Hutton D, Connors L (1999) *History of the Australian Environment Movement*. Cambridge University Press, Melbourne.
58. Hansard Archive 1870–1995. Public Bills of the Session, introduced but not passed (p. 9). Norseman-Esperance Railway Bill. Government Printer, 19 December 1911, Perth.
59. Widell M (2007) Historical evidence for climate instability and environmental catastrophes in northern Syria and the Jazira: the chronicle of Michael the Syrian. *Environment and History* **13**(1), 47–70. doi:10.3197/096734007779748255
60. Allison HE, Hobbs RJ (2004) Resilience, adaptive capacity, and the 'lock-in trap' of the Western Australian agricultural region. *Ecology and Society* **9**(1), 3. doi:10.5751/ES-00641-090103
61. Burvill GH (1988) *The Soils of the Salmon Gums District, Western Australia*. Technical Bulletin 77. WA Department of Agriculture and Food, Perth.
62. Esperance lands and salt. *Western Mail*, 17 August 1917, p. 28.
63. Paterson JW (1917) In *Report of the Royal Commission on Mallee Belt and Esperance Lands*. WA Department of Agriculture, Perth.
64. Royal Commission on the Question of the Construction of a Railway to Pinnaroo, 1905.
65. Esperance Lands Royal Commission: Settlers examined. Experiences in the mallee belt. *West Australian*, 17 January 1917.
66. 66. 'X' [Anon.] (1923) Esperance and eastward. An historic province. Present and future. *Western Mail*, 26 July, p. 2.
67. Kent A (2015) Retired farmer and dam-sinking contractor from Salmon Gums. Interviewed by Nicole Chalmer, Esperance, 7 October.
68. Photographer HOLT. EMA P4158, Battye Library, Perth.
69. *Western Mail*, 17 August 1917, p. 28.
70. Comments by Mitchell and Ministerial Reply. *Sunday Times*, 16 February 1930.
71. Bolton GC (1986) Mitchell, Sir James (1866–1951). *Australian Dictionary of Biography*. Melbourne University Press, Melbourne.
72. Campbell D (1995) Scenes from fifty-two years of farming at Scadden. In *Faith, Hope and Reality: Esperance 1885–1995*. (Coord. P Blumann). Esperance Shire Council.
73. Report of the Registrar General on the Vital Statistics of Western Australia for the Year ended 31st December (1896–1960).
74. Murray D (1985) Agricultural settlement experiments in Western Australia: the Esperance sandplain, 1956–1975. *Rural Systems* **3**(1), 41–62.
75. Department of Agriculture and Food (n.d.) *Dryland Salinity in Western Australia: An Introduction*. WA Department of Agriculture and Food, Perth.
76. Seymour M, Burgess P (2000) *Agriculture Western Australia: Esperance Manual*. Agriculture WA, Esperance.
77. Verboom WH, Pate JS (2006) Bioengineering of soil profiles in semiarid ecosystems: the 'phytotarium' concept. A review. *Plant and Soil* **289**(1–2), 71–102. doi:10.1007/s11104-006-9073-8

9
Why change the land use when you can change the landscape?

> To the Aboriginal people 'good country' is clean country well burned and part of the long-term management cycle ... to keep land food productive for people and animals ... For the Anglo/European [farmers as food producers] good country is clean country also, but clean of trees and 'scrub', cleared land is good land – it produces pasture for the 'good' animals; it provides ground for the 'good' plants.[1]

Perceptions of good land management are dependent on human cultural beliefs as to what is good food.[2] Finding the reasons for the determined efforts to transform nature and landscapes into 'good lands' has been a constant theme throughout this book. As previously discussed, since Aboriginal people throughout Australia depended upon, utilised and managed a wide range of produce from the landscape, there were a variety of 'good lands' manifestations. After at least 50 000 years of adapting to Australia, ecoagriculture passed on through intergenerational knowledge, learning and education, meant most ecosystems were co-evolved with Aboriginal social ecological systems (SES). The food, economic and cultural beliefs of European settlers had also been formed over thousands of years by their adaptive interrelationships with European and British nature as SES. Therefore, to European invaders who became settlers and farmers, the domesticated animals and plants from home were the 'proper, civilised' foods. With an export-oriented agricultural economy, food was not only to eat but also represented a commodity to sell for money. It needed to fit into what Europe, Britain and other markets wanted.[3] From the colonial settler viewpoint, Australian environments and ecosystems would need drastic remodelling to create suitable Old World landscape conditions in which to grow their 'proper, civilised' food.

In Western Australia and the Esperance bioregion, around 92% of all land cleared and developed for agriculture took place during the period from 1953 to 1982. There was a driving state government developmentalist desire to expand agricultural commodity exports, consolidate 'civilisation' and create a Europeanised useful landscape through settling and transforming the bushlands.[4,5] After World War II, an increased rate of change was made possible with new technologies in machinery, fertilisers and chemicals, many stemming from wartime chemical weapons research and development.[6]

Visionaries such as E.F. Smart extolled, in a government-sponsored pamphlet, the passionate desire to turn what was termed the useless 'wilderness wastelands' of the Western Australian sandplains into new agriculturally productive and populated regions.[7,8,i] There was overwhelming evidence by this time that clearing was causing salinisation and that

---
i   Bellanta explains that any unalienated and 'unimproved' crown lands were termed 'wastelands'.

numerous soils were naturally saline throughout many areas of south-western Australia.[9,10] However, there was also a prevailing belief that 'modern' technology could conquer all such setbacks and allow the yeoman vision to continue. Environmental historian Andrea Gaynor describes how:

> ... the older vision of a great agricultural state of bold yeoman farmers also persisted, and where scientists counselled caution in clearing the land for development they were frequently ignored by the public, aspiring farmers and the incumbent politicians alike.[11]

This developmentalism ideology was prevalent throughout Australia since first European settlement, and began in Western Australia when the Swan River Colony was founded on 12 August 1829. As part of the ceremony, Helen Dance officially struck the first symbolic blow against nature, when she drove an axe into a tree on the banks of the Swan River. Developmentalism reached its peak when the Esperance bioregion sandplain became the last huge area of southern Australia to be devegetated and its soils re-engineered as part of this develop-at-all-costs vision.[5] Existing Conditional Purchase (CP) schemes and new policies of encouraging private investment from overseas was used to develop this vast area. It pushed most of the unique ecosystems – ranging from the visible mallee heath and sandplain kwongan plant and animal ecosystems, including the myriad unseen soil ecosystems – to the verge of extinction, before cultural perceptions of agriculture as the only legitimate land use were challenged in the 1980s and 1990s.[12]

Industrialising agriculture

Forms of agricultural industrialisation have occurred throughout human food production history and seem linked to civilisation collapse. Settled civilisations, including Babylon and the Roman Empire, relied upon forms of industrialised agriculture to feed their erupting populations. Highly organised farming systems that produced bulk grains grown in drier country, and efficient methods of food distribution, were essential for their growth-based economies. Eventually, cycles of intensification to feed erupting overpopulation, degraded soils and drought-stricken lands triggered collapses or declines in most of these civilisations as they ran out of food and water.[13] These historical ecological events should be a serious warning to Australia and other countries that are pushing the limits of production, especially in semi-arid lands. Aridification can rapidly increase with unsympathetic land management systems that destabilise hydrological cycles and bioprecipitation when perennial vegetation and water infiltrating soils are destroyed.

Agricultural industrialisation developed during the 19th century in Great Britain as other industries industrialised and populations grew and gravitated towards cities. The previous biodiverse yeoman farms largely disappeared with the farm specialisation and mechanisation trend.[14] Britain lost food self-sufficiency and relied upon agricultural produce from newly colonised lands, which were already following the industrial template and producing bulk food commodities for international markets. Agricultural industrialisation was supported when steamships, refrigeration and freezing developed, allowing exporting and importing of bulk food commodities like grain and frozen meat, by the second half of the 19th century.[3]

One of the greatest vulnerabilities of industrialised agricultural systems is the reliance on very few species and varieties of plants and animals. In the Esperance bioregion, wheat cropping is vulnerable with 85–90% planted to one variety, Mace.[15] Worldwide, over 7000 food species are eaten but only a few animals (chickens, pigs, goats, sheep and cattle) and few varieties of rice, wheat and maize account for 60% of calories and 54% of protein consumed.[14] For instance, the worldwide outbreak of an influenza A virus (African swine fever), which affects pigs, has put the livelihoods and food security of many regions and nations at risk; for example, China's pig population halved in 2019.[16]

Perhaps the most worrying characteristic of industrialised agriculture is that criteria for success are firmly based upon economic sustainability imperatives, rather than long-term social-ecological sustainability. This is reflected in the language used by agricultural institutions, consultants and farmers themselves to describe farm businesses – productivity, maximisation, yield per hectare, dollars per hectare, return on investment, gross margins – never a mention of environment, communities or happiness! Benchmarking success is considered only in terms of economics. Tracey Clunies-Ross and Nicholas Hildyard contend that over the last 40 years, 'agriculture in the industrialised countries has undergone a further revolution' with new development, technologies and commodification:

> The aim by farmers and agricultural scientists was to make agricultural production systems more efficient by reducing the number of species, varieties within species and phenotypes within varieties to the minimum and using technology such as mechanisation, pesticides and manufactured fertiliser to modify environmental conditions to suit these 'improved' plants and animals.[17]

Basim Saifi and Lars Drake agree, and consider these processes as driving industrialised intensification even more.[18] The introduction of genetically modified species has allowed multinational chemical companies such as Monsanto (now owned by Bayer) to deepen their influence on world food security through vertical integration. Their patented genetically modified crop varieties depend upon their configured chemicals for growth and production, and these companies also own the seed rights. Farmers can no longer hold seed back for replanting but must pay for new seed every season.[19] The interactions and connections between the local community, agriculture and nature are weakened, as are feedbacks from social-ecological restraints.

Juan Infante-Amate and Manuel de Molina belong to a multi-country research team that is comparing the true efficiency of industrialised agriculture with that of traditional methods.[20] Olive production is a completed case study. Industrialisation has relegated Spanish agriculture to growing bulk commodified food as raw materials for industrial production processes. Thousands of hectares of monoculture olive farms comprise rows of olive trees on water/fertiliser drip lines, with bare ground beneath.[21] Previously, Spanish olives were grown as low-input biodiverse agro-forestry systems with cattle, pigs, sheep and poultry grazing grasses and fallen olives underneath the trees. Damaged trees provided fuel for cooking and heating and the system featured a variety of other plant foods grown in healthy microclimates and soils. These systems fed families and local communities self-sufficiently. Modern commodity olive orchards are a resource-intensive sector in which the total energy used for production is higher than the food energy produced.[21] In terms of energy input, energy output and long-term sustainability of a SES, multifunctional traditional methods use up to five times less energy than modern industrialised production and are demonstrably more resilient and sustainable.[20]

The South Australian wheat industry from 1869 onwards was Australia's first industrialised agricultural system. It was established in arid country including semi-desert, well beyond sustainable environmental and rainfall stability. Wheat growers became so specialised that they could not feed themselves without cash money earned from the crop. This was a farming SES that mirrored Industrial Age farm systems already evident in Europe and the US. Historian Donald Meinig observed 'It was geared towards mass production of a single [export] commodity within a world-wide system of regional specialisation'.[22] This unremitting desire for industrialised wheat growing illustrates another weakness of economic productivist agricultural policies later embodied in the Great Clearing of Western Australia – the certainty that production is worth doing even if long-term land degradation and unsustainability are inevitable.

After periods of extreme drought, crop failures and devastating soil degradation, the South Australian government tasked its Surveyor General, George Goyder, with developing a safe boundary for wheat growing. He developed an east–west boundary line (Goyder's Line) of 250 mm annual rainfall, beyond which rainfall was too unreliable.[22] With modern minimal tillage systems, cropping was allowed to shift north of this line, but climate change-attributed drought is forcing it south again.

This industrialised system became a model for intensive cropping throughout Australia but took until after World War II to accelerate throughout Western Australia.[23] After the establishment of Grass Patch Farm in 1894, landscape changes were slow-paced as new information and research was incorporated. This adaptive approach allowed the development of a locally suited ecoagriculture that retained some biodiversity. Then wool prices escalated in the 1920s and the push for faster development became the socio-political aim.[24] After World War II, the Western Australian sandplain from Geraldton to Esperance became the new area of excitement. Post-war Esperance farmer Nils Blumann described how early European settler beliefs still persisted, convinced of their moral obligation to transform 'worthless wastelands' into farmlands as evidence of ownership, progress and civilisation in Australia, 'lest people from other countries lay claim to our land'.[25]

Defeating fertility limits

The Western Australian sandplain re-engineering was based on permanent pasture with livestock systems. Subterranean clovers were inoculated with nitrogen-fixing bacteria and planted to provide a fertile base upon which to build soil fertility.[26,ii] Grain growing was

---

ii   Legumes, including serradellas and clovers, fix atmospheric nitrogen into their roots, aided by species-specific symbiotic bacteria.

always the preferred enterprise but couldn't occur until soil nitrogen and organic matter were sufficiently improved by pasture legumes and grasses.[27,iii] Successful pasture establishment required significant nutrient inputs, as F.W. Bow from Esperance showed in 1912, when he successfully planted Italian ryegrass (*Lolium multiflorum*) and a clover, King Island melilot (*Melilotus indicus*), using 112 kg/ha of superphosphate.[28] Western Australia was using artificial fertiliser during the early 1900s, ahead of the UK, where regular use only started in the 1920s. The slower uptake in the UK may be a reflection of greater natural soil fertility and a long history of yeoman SES recycling nutrients.

Pasture researcher C.M. Donald believes improved pasture and livestock systems represented the greatest positive environmental change in Australian agriculture since European settlement. Soil fertility dramatically increased as nitrogen and phosphorus levels increased, and organic matter added carbon to soils. Significantly more livestock could be carried on these improved fertilised pastures compared to native pastures.[29] As areas of improved pastures increased (and native pastures declined), livestock numbers rose linearly up to 100% throughout Australia.

The Department of Agriculture ran pasture fertilising trials in 1949 at Esperance Downs Research Station (EDRS) in collaboration with farmer Alf Button at his sandy Shark Lake property near Esperance, to determine plant nutrients needed for bioengineering the sandplains. They found that sandplain soils were extremely deficient in the macronutrients phosphorus, nitrogen and potassium. Phosphorus levels of around 0.01% were well below the 0.05% needed by introduced plants. The micronutrients cobalt and copper, essential for ruminant livestock and clover nodulation, and copper, molybdenum, sulphur and zinc for plants, were also deficient.[30,31] Further deficiencies have since become known, including selenium for animals, and manganese and boron for introduced plants.[32] Agricultural researchers came to understand German chemist Justus von Leiberg's 1840 concept – his 'Law of the Minimum' – that plants grow in direct proportion to their supply of nutrients, and that a deficiency of any one prevented or limited growth.[33] These intrinsic soil deficiencies make economic growing of broadacre organic foods, that precludes adding artificially produced nutrients, unfeasible – all farmed plants and animals need these essential nutrients.

Agro-systems in the Esperance bioregion during the 1940s and 1950s were still based on the more fertile soils of the mallee and its intrusions in the Esperance sandplain. Albert Kent, who grew up on a farm at Dalyup, described how trace element deficiencies kept most Esperance sandplain farms centred around the fertile patches of the original 'First class' soils.[34] The causes of coastal disease – lack of cobalt, copper and other trace elements – were identified in 1934 by the Commonwealth Scientific and Industrial Research Organisation (CSIRO) Animal Production Division in South Australia, in collaboration with Eric Underwood from the Western Australian Department of Agriculture. Coastal disease could be prevented by including these minerals when topdressing superphosphate. Yet ruminant livestock were still confined to the fertile patches, and farmers who owned sandplain and mallee farms practised transhumance.[35,iv] The low uptake of this new knowledge was likely

---

iii Wind erosion is always an issue on sandplain soils.
iv McKay describes the exciting discovery of how to prevent 'coastal disease'.

linked to the deepening economic depression and farm abandonment, along with inefficient communication networks in agricultural circles.[36]

The Great Clearing: 'they weren't doing anything with it anyway!'[37]

The 1938 Honorary Royal Commission on Light Lands and Poison-Infested Lands was appointed to determine how best to promote the development of the sandplains, these 'light and poison infested lands'.[38] Its report favoured development, noting that up to two sheep per acre were carried on established farms and the climate was extremely reliable.[39] This decision on light lands covered much of Western Australia's unique kwongan and mallee-heath ecosystems. It resulted in a dubious record – the largest area of native vegetation in the world to be cleared in so short a time span.[40]

The Esperance Land Development Committee, which included local farmers, a land agent and the local manager of the Bank of New South Wales, was formed in 1947. The committee published articles promoting the successes of pasture establishment and grazing trials, and persuaded the Minister for Agriculture, G. Wood, to visit in 1948 and view the trials for himself. He was impressed enough to support the 1949 establishment of the Esperance Downs Research Station in the sandplain at Gibson, 29 km north of Esperance.[41]

After the pasture establishment methodology was developed, the R & I Bank recommended government financial support for its establishment to produce meat and wool on the Esperance Downs. The region was declared a special settlement area in 1952, with 2500 acres (1011 ha) assumed to be the viable block size for a family farm. The land was granted under CP conditions that restricted applicants to 2500 acres maximum, to be lived on by the selector or a chosen agent within one year of selection date and to remain the main residence for five years; payments to be made in accordance with the terms of Section 47 of the *Land Act 1933*; properties were to be fenced in accordance with *Land Act* requirements; after the first year, for four years 250 acres (101 ha) per year of pasture were to be developed; and a water supply was to be established as required under the *Land Act*.[42]

In the 1960s, slow horse- and tractor-powered clearing was replaced by Caterpillar D7s and large-gauge anchor chains that rapidly cleared huge swathes of the sandplain vegetation.[43] The landscape-wide soil surveys needed to provide a large baseline for determining land capabilities had not been kept up. Only a small-scale survey on the soils of the Esperance Downs Research Station was completed by 1950.[44] In the 1970s and 1980s, CSIRO was involved in land system style assessments of Western Australia's south coast.[45] However, none of these land system surveys were extensive enough, nor the resulting maps detailed enough, to allow fine-scale objective assessments for determining true land capabilities.

Those dissenting against the wide-scale ecosystem destruction taking place were not listened to, and no record is found of political or bureaucratic concerns about the consequences, except that the soils should not be bare fallowed due to their susceptibility to wind erosion.[46] Salinity potential was never explored as had been done previously by Dr Teakle for the mallee, and it was not until 1985 that a localised assessment was undertaken for the proposed opening of the Beaumont area north-east of Condingup. There was a horrifying ignorance of the hydrology, soil types, and distribution and properties of an irreplaceable natural resource, the topsoils of Western Australia. Nothing had been learnt from previous mistakes when clearing mallee lands. Economic, political and social pressures to re-engineer the sandplains outweighed caution and science.

In 1990 Brian Purdie was appointed by the Western Australian Department of Agriculture to standardise the soil-landscape mapping program methods and outputs. He introduced a 'nested hierarchy of soil-landscape mapping units', which also allowed a relatively seamless inclusion of existing data.[47] This regional mapping program was continued throughout the 1990s and by 2003 was completed to the level of a linked digital coverage, with new information improving the accuracy as an ongoing process. Unfortunately, the thoroughness with which landscapes were cleared and bioengineered, means that there is no true baseline. The maps can only ever be retrospective in their use, and in attempts to deal with the escalating problems caused by overclearing.[48]

By 1954, 36 farmers had developed about 8000 ha out of the total released of 101 215 ha. This was considered too slow by the state government, which appointed The Esperance Downs Development Advisory Committee to investigate ways of speeding up the process.[24,49] Its report found rapid development was hindered because most applicants granted land had inadequate capital and needed outside income to survive the time lag before pastures became established enough for livestock income. Speeding up development would involve some form of government assistance or a high capital private enterprise scheme.

In the late 1940s a syndicate of Americans, including film and television stars including Art Linkletter and Ann Sothern, led by Allen Chase, learnt about Esperance from F.J.S. Wise, then Administrator of the Northern Territory, who had been told about the district's potential by the visiting Western Australian Minister for Lands and Agriculture, E.K. Hoar. The Americans had just lost considerable money in an enormous failed rice-growing venture at Humpty Doo in the Northern Territory, managed by William Gunn, Commonwealth Bank director and friend of the deputy Prime Minister John McEwan.[50] The Americans signed a development agreement with the Western Australian state government on 19 November 1956 which allowed their company, Esperance Plains (Aust.) Pty Ltd, to acquire 600 000 ha of land at around 35 pence/ha plus survey costs. They would clear and develop the land, plant pasture and on-sell half of it as 800 ha blocks with a house and fencing, with first preference to Australian buyers.

The syndicate regrettably retained faith in William Gunn, choosing to ignore advisor Merton Love, who recommended following the land development techniques proven by locals and the Department of Agriculture at Esperance Downs Research Station. Instead, the syndicate followed the recommended 'shortcuts' advised by William Gunn and Dr Moule, a veterinarian from Queensland. Jesse Skoss, employed as the on-ground project overseer, also advised the syndicate to follow local knowledge – and was also ignored.[51,52] Ross White, Government Surveyor for this project, later stated that the development procedures were 'a shambles' that failed spectacularly as the syndicate attempted to compress a proven three-year process into one year.[53,54]

By 1959 the Chase Syndicate investors had lost around £500 000 and were completely disillusioned with their development team. Their 100% tax-deductible investment (they could claim tax deductions within the US tax system for new land development overseas) had not returned any income. Art Linkletter persisted with his 4000 ha, suspended his initial scepticism about the value of local knowledge, dumped William Gunn and hired John Hagon as manager. In 1968 he acknowledged that 'John Hagon could very well be the basic reason for my success in Australia'. Hagon used the proven procedure which, though taking three to four years, established high-quality annual pastures.[50] Most land developers established annual pastures with subterranean clover and rye grass, and applied superphosphate, copper, zinc and later molybdenum. Stock were introduced in autumn of year 3 or 4, after which cobalt sulphate would be included.[51] A few forward thinkers like Jesse Skoss included the perennial grasses couch grass, kikuyu and phalaris to provide summer/autumn feed, with a second, deliberate aim of preventing salinity.[52]

After three years of failure to establish significant pasture, the government ruled that the syndicate was in default. In 1959, the company sold nearly 24 000 ha of its undeveloped land and returned 111 903 ha to the state. There were numerous heated parliamentary and public media demands to return the land still held. Art Linkletter irreverently noted in his book *Linkletter Downunder,* 'it is a cardinal rule of politics never to admit to having had made a mistake', so Allen Chase was allowed to introduce a second group of investors. Amfac Inc., a Hawaiian-based sugar and pineapple conglomerate, teamed with the Chase Manhattan Bank (unrelated to Allen Chase), venture capitalist J.H. Whitney & Co. and Australian pastoral house Elder Smith & Co to form the Esperance Land & Development Co.[50] With state government approval and special legislation – the *Esperance Lands Agreement Act 1960* – the company successfully developed the sandplain country as originally envisaged.[55] There were specifications in the Act that required settlers to be at least 50% Australian, with the rest being American or European, and each settler was allotted a single holding. The Americans eventually sold out to National Mutual Life in 1988 and the company was renamed Esperance Rural Properties. National Mutual Life was taken over in 1995 by AXA, a French life insurance company, and by 2002 all the rural assets had been sold to individual farmers and to corporate and overseas investors.[54]

In the meantime, the state government released the land returned to it from the Chase scheme as CP blocks of up to 1800 acres (728 ha) east of Esperance and 1600 acres (647 ha) to the west. The purchase price of $800 seemed cheap, but it cost another $20 per acre to establish pasture. Many settlers were undercapitalised and needed income from other employment, to succeed. Luckily the Esperance Land & Development Co. employed a number of settlers for its own development, but the block sizes eventually proved too small for a decent living as wheat, wool and beef prices fell, and input costs went up. As the first Esperance Agricultural Department Manager, Geoff Grewar, explained, 'a lot of early settlers in the 1960s and 1970s had to leave because the input costs beat them'. With declining returns, banks foreclosed and despite endless work the future was grim, especially when 'the wife gets sick of living in a shed' and can see no lifestyle resolution.[56] The role of farming women is extremely underplayed in literature about farming business development, including in present times. That is why I always refer to myself or other farm women as farmers (definitely not as 'farmers wives'!)

The earlier, more limited clearing methods using horses or tractors were more environmentally friendly, as paddock boundary windbreaks containing the large trees and bush were left.[57] The 1960 *Esperance Lands Agreement Act* had agreed to leave natural vegetation where possible, 'The Co. shall as far as practicable preserve natural growth and trees which are not harmful to agriculture or pasture and encourage settlers in the planting of suitable trees'. Unfortunately, contradictory government conditions meant the Esperance Land & Development Co. did not have to comply, and it strove to clear as much land as possible. The D7 dozers which arrived in 1960, had 145 drawbar horsepower and could pull a 200 m length of 5 cm anchor chain between them at a speed of 5 km h^{-1}. Clearing a swathe 75–80 m wide meant on average 250 ha a day was flattened. Big trees were furiously attacked with a pusher bar mounted above the dozer blade, so that the chain could continue unhindered on its destructive path.[36]

Despite Jesse Skoss' promotion of perennial grasses to alleviate salinity, little official thought was given to the long-term consequences of the wholesale destruction, especially in causing salinity, wind and water erosion. There appeared to be absolutely no understanding that local and regional climate could be affected.[58,v] There was no encouragement to leave uncleared hilltops, natural drainage lines, riparian zones along creeks and rivers, yate and paperbark swamps and wetlands, and there appears to have been no consideration given to wildlife habitat at all.[59] Though the destructive behaviour was mostly accepted, lifelong conservationist and farmer Jack Ewert decried the attitudes of the time:

> It absolutely horrified me in the 1960s when they started opening up this area on a broad scale ... they would just flatten everything. The paperbark swamps were all wiped out. It nearly drove me crazy. You could never get through to these people ... Some of the required knowledge has been available since the 1940s.[60]

He believed it was a disastrous situation and future farmers would have to deal with the impacts. The problems rapidly appeared, including regular flooding of the Lake Warden Ramsar-listed wetlands which has left no breeding and feeding beach margins for migratory wading birds (coming from as far as Siberia, China and Malaysia); rising salinity impacting farmland and non-farmed ecosystems; and catastrophic flooding events that wash out roads, bridges and downstream farms, and are causing sedimentation of estuaries and inlets.[48]

The removal of areas containing poisonous *Gastrolobium* plants was really the only justification for the wholesale bush clearing. Poison had been the single largest cause of livestock death in the bioregion since statistics were first recorded in 1896.[61] The presence of these plants was a major deterrent to keeping windbreaks and other remnant vegetation patches on gravelly soils, along creeklines, around granite outcrops and in the strip of transitional country between the mallee and sandplain. Most farms cleared riparian vegetation where *Gastrolobium* thrived, turning their creeklines into degraded salty drains.[62] But some farmers have taken a different approach – instead of clearing everything, Orleans

---

v Farmer comments associated clearing with reduced rainfall throughout the district.

Bay farmers Geoff and Mary Hoggart have fenced off their attractive bushy creekline containing *Gastrolobium* plants.

The rampant developmentalism typical of Australian agricultural development, is thoroughly illustrated by the well-financed US Bitelle Institute's proposal to develop 2 025 000 ha from Israelite Bay to Balladonia. Along with several locals, the company formed the West Australian Development Corporation and in 1967 applied to the state government for a licence to develop the region. A major proposal was to 'alter the climate and improve rainfall' by bituminising several hundred square kilometres surrounding spectacular Israelite Bay, where whales give birth. If allowed, this would have been an environmental disaster leading to severe aridification of millions of hectares. Fortunately, the scheme was locally unpopular, with strong feelings about the perceived 'land grab' by Americans. There was enough hostility that the Brand government (1959–1971) rejected the licence, not because of environmental concerns, but as politically risky.[63]

The 1970s was a period of converging challenges for sandplain farmers. Throughout the region the clover-based pastures were decimated by a fungal disease, clover scorch (*Kabatiella caulivora* (Kirchn)), causing massive livestock feed shortages and wind erosion.[64] In 1969 a worldwide oversupply of wheat caused the introduction of national quotas. As industry latecomers, sandplain wheat growers were denied quotas, which led to economic hardship.[65] Wool prices crashed, and in 1976 cattle prices plummeted following oversupply in the US. Luckily for farmers, the superphosphate subsidy introduced in 1941 kept its price at $AU15 per tonne, allowing continued applications and no interruptions to development.[24,66]

Pressure to release new land continued, and in 1985 surveys in the Mt Beaumont, Mt Ridley and Cascades areas were completed. The Shire of Esperance and others were eager for new CP land releases to aid shire income from rates, population growth, road development and business stimulation. An advisory board – the Rural and Agricultural Industries Commission – set up to review options was made up entirely of farmers. Naturally, they proposed developing millions of hectares on the southern coast.[49] Caution and scientific concerns were again overruled by the developmentalist vision. As Ron Richards describes in *Diversity or Dust*, 'the unquestioned default use for unalienated [crown] land was agricultural development unless proved otherwise'.[67] Richards also relates how the Department of Conservation and Environment and Environmental Protection Agency representatives were weak, ineffective members of the 1970s–1980s interdepartmental Working Group on Land Release (Departments of Lands, Agriculture, Fisheries and Wildlife were represented). They never demanded surveys of the fauna and flora or expressed concerns about the science-based warnings of salinity risks in the land release areas.[68]

Rise of conservation in the 1980s

Geographer Arthur Conacher claims that no other region in the world as large as the Western Australian wheatbelt has been cleared of its native vegetation so rapidly, supported by all political factions.[69] The economic extractivist philosophy overruled environmental concerns and future generations will have to deal with the problems.[70] Such attitudes reflect a common human weakness of discounting the future. The attitude is not unique to Australia. When the Esperance Agriculture Department's regional hydrologist was seconded to China to impart research findings and warnings about salinity and hydrology, he found his developmentalist Chinese colleagues equally unconcerned about taking a conservative approach as 'future generations of scientists would solve any problems'.[48]

During the 1980s salinisation was clearly expanding throughout the earlier settled wheatbelt and rapidly appearing on Esperance sandplain farms. Farmer Marg Agnew related, 'By the mid-1980s I could see that the land on the home farm was very slowly going out of production on low-lying areas and around a saltlake'.[71] In response, the Department of Agriculture recommended retaining 10% of native vegetation (based on guesswork) per property as a condition for new CP blocks. Little effort was expended on checking compliance despite the *Soil and Land Conservation Act 1945* (WA) giving the power to do so.[72]

The recommendation was absurd. There were no coordinated objectives, no idea about types of bush to be retained or their distribution on farms and catchments, and without fencing requirements the remnant bush rapidly degraded under continuous livestock pressure.[73,74] During the 1990s, the 10% requirement was raised to 20%, uncleared in large interconnected blocks and preferably fenced from livestock. This was not based on research; it was merely a socially and politically acceptable change. With revegetation help given only through national Landcare grants, realistically covering costs was unlikely. Landcare became dominated by ideological purists who would only fund local native vegetation regeneration, which was impractical in many situations, especially when fertiliser applications kept soil fertility higher than native vegetation was adapted to grow in.[17]

The elephant in the room was and still is – how much bush is enough? Maybe no-one really dares to acknowledge officially or publicly the inconvenient truth. Research clearly confirms a minimum 90% of a catchment needs to grow perennial vegetation (grasses, legumes, shrubs and trees) to prevent and reverse rising water tables and salinity.[40,75] If this was actually done Australia-wide, there would also be wonderful climatic, rainfall and carbon sequestration outcomes.

Forms of state Landcare have existed in Australia since the 1930s – often farmer-driven in response to local problems caused by neighbouring unsustainable farming practices. These could be actioned against in Western Australia by the *Soil and Land Conservation Act 1945*, amended by the *Soil and Land Conservation Act 1984*.[72] A national approach to Landcare began in 1989 when Rick Farley from the National Farmers Federation and Phillip Toyne of the Australian Conservation Foundation jointly convinced the federal Hawke Labor government and the Opposition to take a bipartisan approach to fund a $320 million National Landcare Program. This program helped change many attitudes via education and interactive projects to address issues on ground, especially when environmental concerns could be meshed with production goals.[76]

Though achieving many project outcomes, Landcare has been unable to significantly address the landscape-wide problems throughout Australia. The primary reasons for degradation, including clearing of native ecosystems, loss of biodiversity, lack of landscape-wide perennials, economically driven developmentalism and farm income needs, are unadmitted and unaddressed. The appetite for environmental sustainability action throughout

Western Australia and Australia is secondary to production goals, as farming costs and economic drivers have escalated, to the point where many past revegetation projects, never legally protected by Landcare beyond 10 years, are now being cleared. In Western Australia, salinisation continues in the wheatbelt and Esperance bioregion. Future salinity outcomes are predicted to be worst to the west of the Esperance–Norseman Highway, and eventually will consume lands east of Esperance as well.[48,77,vi] Solving these problems requires a radical cultural shift into supporting perennial-based farming systems helped by cultural food flexibility markets, sympathetic economic systems, communities and government.

In 1984 the Australian Conservation Foundation supported the release of the Working Group on Land Release report, *Diversity or Dust*. This report described the consequences of clearing Western Australian crown land for agriculture, and the consistent history of little or no environmental, social or economic planning.[78] The 'last straw' in 1980 for conservationists and scientists was an announcement by the Minister for Lands, David Wordsworth, that 100 000 ha of south coastal crown land would be released annually for agricultural development until 3.1 million ha was reached. He attempted to quell protests by not to allowing public comment on the Rural and Allied Industries Council's report on new land release, and threw open 50 blocks annually.[79] This was despite the appearance of salinity after clearing in previous land releases and existing wind erosion issues, with vast sand drift areas caused by ploughing and overstocking with sheep.[80] Local farmers I spoke to confirmed that in the 1980s species already declared rare and endangered in the south-west (such as malleefowl, woylies, ringtail possums and tammar wallabies) were still living in Esperance bioregion bushlands but their habitat was being destroyed.[81] Protection of a species is pointless unless their home habitats are also protected.[82]

In 1983 C.V. Malcolm published an extremely detailed Agricultural Department bulletin discussing the historical causes and effects of salinity. It was so comprehensive that it left absolutely no excuse for ignorance about the impacts of land clearing.[83] This document possibly contributed to Premier Brian Burke and Cabinet deciding to approve an amendment to the *Soil and Land Conservation Act 1984*, that disallowed clearing of native vegetation without an assessment and issuing of a clearing permit by the Department of Agriculture. In April 1985, a moratorium on new land release and clearing was passed as the *Environmental Protection Act 1986*, reinforced by the *Native Vegetation Protection Act*. At last, land clearing in Western Australia was put on indefinite hold.[84,85] Other states are still to catch up.

Conserving native ecosystems or revegetating for perennial landscapes on farms is still not clearly recognised as an ecoagricultural problem, in a society which accepts industrialised agriculture for cheap food. Instead, the onus has been on farmers to be ecologically literate, recognise the need and bear the cost of rectifying damage often caused under CP guidelines. Those concerned enough have fenced and revegetated for years, having an empathy for nature and aesthetics, and taking a longer-term view than immediate economic returns. There is still no easy way to protect farm revegetation in perpetuity. The next property owner can destroy 20 or 30 years of revegetation work with impunity because it is not the

---

vi If trees and native vegetation are to be used to reverse salinity, over 90% of the catchment would need replanting.

original native vegetation.[vii] Pragmatic farmers who are intent on meeting yearly financial obligations may disagree that bush makes their individual farm businesses more profitable. Even though new research is showing otherwise, it is not being communicated effectively within the agro-political arena. There is sometimes underlying resentment, for unsustainable developmentalism was relentlessly pushed in the past by politicians and other members of society. Farming communities alone will be unable to bear the costs of revegetating Australia – it is a project for all Australians.

Endnotes

1. Mears G (2011) *Foal's Bread*. Allen and Unwin, Sydney.
2. Anderson EN (2005) *Everyone Eats*. New York University Press, New York.
3. Ponting C (2007) *A New Green History of the World: The Environment and the Collapse of Great Civilizations*. Random House, New York.
4. Bradshaw CJA (2012) Little left to lose: deforestation and forest degradation in Australia since European colonization. *Journal of Plant Ecology* **5**(1), 109–120. doi:10.1093/jpe/rtr038
5. Layman L (1982) Development ideology in Western Australia, 1933–1965. *Australian Historical Studies* **20**(79), 234–260. doi:10.1080/10314618208595682
6. Shiva V (2016) *The Violence of the Green Revolution*. Kentucky University Press, Lexington.
7. Bellanta M (2002) Clearing ground for the New Arcadia: utopia, labour and Environment in 1890s Australia. *Journal of Australian Studies* **26**(72), 13–20. doi:10.1080/14443050209387734
8. Smart EF (1965) *Transformation of Wastelands Western Australia – also Advice to Young Farmers with Small Capital*. State Publisher, Perth.
9. Paterson JW (1917) In *Report of the Royal Commission on Mallee Belt and Esperance Lands*. WA Department of Agriculture, Perth.
10. Burvill GH (1979) *Agriculture in Western Australia: 150 Years of Development and Achievement, 1829–1979*. University of Western Australia Press, Perth.
11. Gaynor A (2002) Looking forward, looking back: toward an environmental history of salinity and erosion in the eastern wheatbelt of Western Australia. In *Country: Visions of Land and People in Western Australia*. (Eds A Gaynor, M Trinca, A Haebich). Western Australia Museum and Lotteries Commission of Western Australia, Perth.
12. Australian Greenhouse Office (2000) *Land Clearing: A Social History*. Technical Report 4, September. National Carbon Accounting System.
13. Diamond J (2005) *Collapse: How Societies Choose to Fail or Succeed*. Penguin, Melbourne.
14. Thrupp LR (2000) Linking agricultural biodiversity and food security: the valuable role of agrobiodiversity for sustainable agriculture. *International Affairs* **76**(2), 265–281. doi:10.1111/1468-2346.00133
15. Daw M (2017) Esperance Port Zone Manager, Co-operative Bulk Handling. Pers. comm.
16. <https://www.cnbc.com/2019/10/03/african-swine-fever-chinas-pig-population-may-drop-by-55percent.html>
17. Clunies-Ross T, Hildyard N (2013) *Politics of Industrial Agriculture*. Routledge, London.
18. Saifi B, Drake L (2008) A coevolutionary model for promoting agricultural sustainability. *Ecological Economics* **65**(1), 24–34. doi:10.1016/j.ecolecon.2007.11.008
19. Robin M-M (2010) *The World According to Monsanto: Pollution, Politics and Power*. New Press, New York.
20. Infante-Amate J, De Molina MG (2013) 'Sustainable de-growth' in agriculture and food: an agro-ecological perspective on Spain's agri-food system (year 2000). *Journal of Cleaner Production* **38**, 27–35. doi:10.1016/j.jclepro.2011.03.018
21. Infante-Amate J, De Molina MG (2013) The socio-ecological transition on a crop scale: the case of olive orchards in southern Spain (1750–2000). *Human Ecology* **41**(6), 961–969. doi:10.1007/s10745-013-9618-4
22. Meinig DW (1962) *On the Margins of the Good Earth*. Rigby, Adelaide.

---

vii When a friend sold her farm, her 25 years of revegetation were destroyed in two years. The new owners cleared all revegetation for cropping.

23. Report of the Registrar General on the Vital Statistics of Western Australia for the Year ended 31st December (1896–1960).
24. Rintoul J (1986) *Esperance Yesterday and Today*. 4th edn. Shire of Esperance, Esperance.
25. Blumann N (1995) Attitudes. In *Faith, Hope and Reality: Esperance 1885–1995*. (Coord. P Blumann). Esperance Shire Council, Esperance.
26. Anon (1949) Esperance News. Farming Experiments. *Kalgoorlie Miner*, 15 June, p. 2.
27. WA Department of Agriculture and Food, Grains Research and Development Corporation (2009) *Managing South Coast Sandplain Soils for Yield and Profit*. Bulletin 4773. Perth.
28. WA Department of Agriculture (1960) *Agriculture on the Esperance Downs*. Bulletin 3080, 1–2.
29. Donald CM (1965) The progress of Australian agriculture and the role of pastures in environmental change. *Australian Journal of Science* 27(7), 187–198.
30. Fletcher C (1995) Alfred George Button, Father of Esperance Agriculture. In *Faith, Hope and Reality: Esperance 1885–1995*. (Coord. P Blumann). Esperance Shire Council, Esperance.
31. McKay A (1976) Cobalt and coast disease. In *Surprise and Enterprise: Fifty Years of Science for Australia*. (Eds F White, D Kimpton) pp. 18–20. CSIRO Publishing, Melbourne.
32. Hamza MA (2008) *Understanding Soil Analysis Data*. Report 327. WA Department of Agriculture and Food, Perth.
33. Loneragan JF (1999) Nutrients: a sparse resource. In *Plants in Action: Adaptation in Nature, Performance in Cultivation*. (Eds BJ Atwell, PE Kriedemann, CG Turnbull) Ch. 16. Macmillan Education, Melbourne.
34. Kent A (2015) Former farmer, Esperance. Interviewed by Nicole Chalmer, 7 October.
35. McKay A (1976) Cobalt and coast disease. In *Surprise and Enterprise: Fifty Years of Science for Australia*. (Eds F. White, D. Kimpton) pp. 18–20. CSIRO Publishing, Melbourne.
36. Campbell D (2006) Scadden. In *Pioneers and Early Settlers of Scadden, 1910–1959*. (Ed. L Bale). Publisher unknown.
37. From one of the old farmers of Esperance, regarding Aboriginal land management. Pers. comm.
38. Grewar G (1995) From farm to corridors of parliament and back. In *Faith, Hope and Reality: Esperance 1885–1995*. (Coord. P Blumann). Esperance Shire Council, Esperance.
39. Western Australia. Honorary Royal Commission on Light Lands and Poison-Infested Lands 1938–2008.
40. Beresford Q (2001) Developmentalism and its environmental legacy: the Western Australia wheatbelt, 1900–1990s. *Australian Journal of Politics and History* 47(3), 403–415. doi:10.1111/1467-8497.00236
41. Department of Environment and Conservation (n.d.) *Historical Forestry Notes: Esperance*.
42. Special settlement area near Esperance. *West Australian*, 10 March 1950, p. 13.
43. Freeman K, Freeman B (1995) The Grass Patch. In *Faith, Hope and Reality: Esperance 1885–1995*. (Coord. P Blumann). Esperance Shire Council, Esperance.
44. Smith ST (1950) Soil survey of the Esperance Downs Research Station. Unpublished report. WA Department of Agriculture. In *Esperance Land Resource Survey*. (Eds TD Overheu, PG Muller, ST Gee, GA Moore). Report 8, 1993. WA Department of Agriculture and Food, Perth.
45. Northcote KH, Bettenay E, Churchward HM, McArthur WM (1967) *Atlas of Australian Soils: Sheet 5, Perth-Albany-Esperance Area with Explanatory Data*. CSIRO Australia and Melbourne University Press, Melbourne.
46. Richards R (1984) Agricultural accountability. In *Diversity or Dust: A Review of Land Clearing Programs in South West Australia*. Australian Conservation Foundation, Perth.
47. Purdie BR, Tille PJ, Schoknecht NR (2004) *Soil-Landscape Mapping in South-Western Australia: Overview of Methodology and Outputs*. WA Department of Agriculture, Perth.
48. Simons J (2013–2017) Esperance Regional Hydrologist. Interviewed by Nicole Chalmer, WA Department of Agriculture and Food Office, Esperance.
49. Six will advise on Esperance Downs. *West Australian*, 20 January 1954, p. 1.
50. Linkletter A (1968) *Linkletter Down Under*. Prentice Hall, New Jersey.
51. Hagon J, Kirwan W (1995) The beginnings of modern Esperance. In *Faith, Hope and Reality: Esperance 1885–1995*. (Coord. P Blumann). Esperance Shire Council, Esperance.
52. Dr Jesse D. Skoss, Palo Verde Plant Research Resource Associates. Letter to Avon at Esperance Museum, 'Regarding the early days at Esperance and the beginning of the Chase Land Development fiasco'. Skoss denies any part in the matter.

53. White R (1995) Surveying for the Esperance pioneers. In *Faith, Hope and Reality: Esperance 1885–1995*. (Coord. P Blumann). Esperance Shire Council, Esperance.
54. Grewar G (2012) Farmer and owner of Thomas River Station. Interviewed by Nicole Chalmer, Coronet Hill, 20 August.
55. Senior B (1995) The coming of the Americans, 1956–1988. In *Faith, Hope and Reality: Esperance 1885–1995*. (Coord. P Blumann). Esperance Shire Council, Esperance.
56. Hockey S (1995) Starting in Esperance in the early 1950s. In *Faith, Hope and Reality: Esperance 1885–1995*. (Coord. P Blumann). Esperance Shire Council, Esperance.
57. Reichstein D (1995) Machines of the modern pioneers. In *Faith, Hope and Reality: Esperance 1885–1995*. (Coord. P Blumann). Esperance Shire Council, Esperance.
58. Kala J, Lyons TJ, Nair US (2011) Numerical simulations of the impacts of land-cover change on cold fronts in south-West Western Australia. *Boundary-Layer Meteorology* **138**, 121–138. doi:10.1007/s10546-010-9547-3
59. Platt J (1995) The development of Esperance landscapes. In *Faith, Hope and Reality: Esperance 1885–1995*. (Coord. P Blumann). Esperance Shire Council, Esperance.
60. Moran R (1995) Jack Ewert: farmer and long-time conservationist. In *Faith, Hope and Reality: Esperance 1885–1995*. (Coord. P Blumann). Esperance Shire Council, Esperance.
61. Report of the Registrar General on the vital statistics of Western Australia for the year ended 31st December (1896–1960).
62. Hoggart M (2014, 2016) Farmer and botanist. Interviewed by Nicole Chalmer, Condingup, 10 November 2014, 15 August 2016.
63. West Australian Development Corporation (n.d.) *Land for Grazing and Farming in Shires of Esperance and Dundas*.
64. Chatel DL, Francis CM (1974) *The Reaction of Varieties of Subterranean Clover to the Clover Scorch Disease (Kabatiella caulivora (Kirchn) Karak) at Three Sites in Western Australia*. Technical Bulletin 25. WA Department of Agriculture and Food, Perth.
65. Wheat quotas: urgent need for quota system. *Beverley Times*, 6 June 1969, p. 4; *Wheat Delivery Quotas Act 1969*.
66. Ryan T (2010) *The Australian Fertilizer Industry: Values and Issues*. Australian Fertilizer Industry Conference 2010. Fertilizer Industry Federation of Australia, Canberra.
67. Richards R (1984) Structure and process. In *Diversity or Dust: A Review of Land Clearing Programs in South West Australia*. Australian Conservation Foundation, Perth.
68. Richards R (1984) The bureaucracy. In *Diversity or Dust: A Review of Land Clearing Programs in South West Australia*. Australian Conservation Foundation, Perth.
69. Conacher A (Ed.) (2001) *Land Degradation: Papers Selected from Contributions to the 6th Meeting of the International Geographical Union's Commission on Land Degradation and Desertification*, Perth, 20–28 September 1999. Kluwer Academic, Dordrecht.
70. Wilk R (2004) The extractive economy: an early phase of the globalisation of diet. *Review: Fernand Braudel Center* **27**(4), 285–306.
71. Agnew M (1995) Farming for the future. In *Faith, Hope and Reality: Esperance 1885–1995*. (Coord. P Blumann). Esperance Shire Council, Esperance.
72. *Soil and Land Conservation Act 1945*. Originally set up at the request of land users – an Act 'relating to the conservation of soil and land resources, and to the mitigation of the effects of erosion, salinity and flooding', it had the power to prevent further clearing of remnant vegetation on private lands.
73. Chalmer NY (n.d.) Pers. obs.
74. Harkness W, Harkness P (2016) Farmers at Beaumont. Interviewed by Nicole Chalmers, Beaumont, 20 October.
75. Ferdowsian R, George R, Lewis F, McFarlane D, Short R, Speed R et al. (1996) The extent of dryland salinity in Western Australia. In *Proceedings of 4th National Conference on Productive Use of Saline Lands and Workshop on the Productive Use and Rehabilitation of Saline Lands*, Albany, 25–30 March, pp. 89–97.
76. Wilson GA (2004) The Australian Landcare movement: towards 'post-productivist' rural governance? *Journal of Rural Studies* **20**, 461–484. doi:10.1016/j.jrurstud.2004.03.002

77. Coyne P, Williamson DR, Thomas JF (n.d.) *Salinity Crisis Action Plan: Western Australia*. Government Printer, Perth.
78. Australian Conservation Foundation, April 1984.
79. Wordsworth D (1995) From Tasmania to Mediterranean Esperance. In *Faith, Hope and Reality: Esperance 1885–1995*. (Coord. P Blumann). Esperance Shire Council, Esperance.
80. Bradby K (2016) CEO of Gondwana Link. Interviewed by Nicole Chalmer, Albany, 26 March.
81. Bertola P, Bertola P (2014) Mallee farmers at Beaumont. Interviewed by Nicole Chalmer, Beaumont, 17 June.
82. Richards R (1984) *Agricultural Accountability*. Australian Conservation Foundation, Perth.
83. Malcolm CV (1983) *Wheatbelt Salinity: A Review of the Salt Land Problem in South Western Australia*. Technical Bulletin 52. WA Department of Agriculture and Food, Perth.
84. Squelch J (2007) Land clearing laws in Western Australia. *Legal Issues in Business* **9**, 1–15.
85. State Records Office of Western Australia (2015) *A Guide to the 1984 State Cabinet Records*.

10
Comparing pathways, past and present: the path chosen may not let you return

> ... the protection of the natural world is an even more fundamental moral and ethical principle than the good of humankind. In other words, the natural world comes first, before everything and everyone else.[1]

Paul Collins is an Australian ex-Catholic priest, historian, environmentalist and ABC presenter. He calls for us to rethink our place in the world and recognise that humanity has not been given the right of dominion over nature, which doesn't exist solely for our use and exploitation. For our survival, nature's needs must come first for nature does not need us, but we need nature.[1] American environmental philosopher and farmer Wendell Berry echoes this call. He believes that agriculture provides a tangible way to alter this viewpoint, through recognising that farms managed responsibly are farmed as ecosystems.[2] The values reflected will ensure very long-term sustainable nature and farm futures. In reality, nature's needs are usually thoughtlessly and deliberately overridden by the economic and cultural values of the greater society that we live in. The production of healthy, tasty and nutritious food in healthy ecosystems has become at the most a side issue. Modern industrial agriculture is very often valued less for food than for its economic, social and geopolitical purposes.[3]

After World War II, throughout Australia entire landscapes were rapidly cleared and re-engineered into standardised industrial modes of agricultural commodity production. World agriculture has continued treading this path of enormous deforestation and dominance by monoculture systems and is increasingly dependent on inputs from somewhere else to produce products to sell elsewhere. William Catton, an environmental economist, describes this as exploiting the 'ghost' resources and economies of other lands and peoples to maintain the fallacy of endless economic growth. Colonial food imports from 'ghost' places, including the colonies of India, Africa, the Americas and Australia, into Great Britain and Europe, allowed population growth and lifestyles well beyond the nations' intrinsic carrying capacities. This model now encompasses agribusiness worldwide.[4,5]

Modern industrial farming systems are run under the 'command and control' paradigm discussed by Carl Holling and co-researchers.[6] As the complexities of nature are reduced in order to increase the predictability and stability of the agro-system in order to achieve maximum economic production, these emergent systems encourage serious environmental problems such as the runaway clearing of forests for cheap timber for the developed world (this includes China and India), and replacing them with industrial agro-systems such as palm oil plantations and soya bean fields.[7]

World humanitarian problems are accelerating as invasion culture continues. Indigenous peoples throughout Asia, Africa and South America are dispossessed of their traditional lands, culture and self-determination, with the false promise of economic growth from the efficiencies of industrialised agricultural exports – just not for them. Indian food and social agriculture researcher Vandana Shiva discusses how Indian agricultural industrialisation, commonly called the Green Revolution, once lauded as a great step forward, is now well documented as threatening India's food security. The past frameworks of co-evolution, ecological and social resilience which had been India's ecoagriculture for thousands of years, have been dismantled. India's current environmental problems parallel many in Australia, including desertification, loss of biodiversity, dependence on monocultures, scarcity of water, depletion of aquifers, increasing salinity and overdependence on chemicals and fertilisers.[8] Agricultural species numbers are being reduced to a fraction of their original biodiversity as new monocultures of soy (*Glycine max*) and oil palms (*Elaeis guineensis*) expand. The accelerating rate of extinctions throughout the world is another powerful indicator that social ecological systems (SES) everywhere have lost any previous sustainability and are collapsing because of modern food producing systems and the societies they feed.

In the Esperance bioregion and similar regions, the three main industrialised farming enterprises are: cropping only of cereals, canola and some legumes such as lupins or peas in vast machine-ready paddocks; mixed farming with cropping and livestock such as sheep (mostly) and/or cattle; and livestock enterprises where sheep or cattle are farmed singularly or together. Our poorly adapted cultural paradigms are reflected in the industrial agriculture fixation on fast-growing early succession annual crop plants like wheat, in environments that had evolved with perennial plants that maintained hydrology for millennia, and in which Aboriginal SES had thrived for thousands of years.[9,10] The methods employed to maximise net production and returns are described by Brian Walker and David Salt, 'the way we do business is sometimes referred to as … command and control' as it involves controlling or commanding aspects of a system to maximise returns.[11] For instance, annual cropping plants occupy the same ecological niche as the tougher weeds that plague them, requiring a constant battle of extermination – in the past with the plough, but now mainly with chemicals. Monocultures are threatened by nature from all sides – by weeds, fungi, bacteria, insects and other animals that take part in developing ecological complexity. The language used to promote chemicals reflects this embattlement, with names such Terminator, Tigrex, Maverick and Crucial used to encourage purchase. Technological solutions that require costly chemical inputs and new mechanical technologies are normally touted as the way forward, with vast amounts of research and development time put into this battle. Though the agro-systems themselves are ecologically simple, there is an immense complexity of physical, social and economic systems needed to maintain them against nature's efforts towards complexity.

There is a significant reliance on fertilisers bought in from elsewhere, to replace nutrients exported off the farm (1 tonne of wheat removes 23 kg of nitrogen, 3 kg of phosphorus and 4 kg of potassium).[12] Farm families' emotional stress is another feature of industrial agriculture, as economic viability needs prices to cover the cost of production every year and to make a profit, but the general trend is of increasing farm debt.[13] Many Australian farmers carry significant bank debt run on yearly cycles which generally do not consider volatile market and climatic fluctuations. Furthermore, agriculture operates under a hybrid market system in which purely competitive sellers are severely disadvantaged because they are selling into and purchasing from monopoly-structured input and output markets.[15] Factors out of farm control that influence profitability include climate and weather, frequency of good seasons, adaptability to poor seasons, market supply conditions, pest and disease fluctuations, input costs of chemicals, fertilisers, fuel and labour and the emotional and economic effects of adverse political decisions such as the federal government shutdown of the live export cattle industry in 2011 (which has since been ruled as illegal).

Sustainability, resilience and change

There have been numerous definitions of sustainability that depend upon its interpretation in various contexts. The Brundtland Commission report of 1987, *Our Common Future*, defined sustainable development as 'development that meets the needs of the present without compromising the ability of future generations to meet their own needs'. This has become the unquestioned position – that development can be sustainable, meeting anthropomorphic needs without impacting nature.[14] In 1989, economist Herman Daly and several other researchers declared that 'sustainable development' was an 'oxymoron' and its assumptions were highly questionable.[15] Since then, new Sustainable Development Goals were developed by the United Nations in 2015, and reported by the Leadership Council of the Sustainable Development Solutions Network.[16] However, nature was again ascribed no fundamental rights, no parameters were set for meeting its multiple needs, and there was a failure to recognise that without healthy nature as pre-eminent, human needs are ultimately not sustainable at all.[17] Viktoria Spaiser and co-researchers have reviewed these goals and conclude that a comprehensive sustainable development theory has not been achieved, attainment of the goals is unclear and environmental damage treated as an externality is still unaddressed.[18]

The Western Australian Department of Agriculture and Food's aspirations for sustainable development aligns with the Brundtland report's definition of sustainability. Its 2013 *Report Card on Sustainable Natural Resource Use in Agriculture* fails to mention nature as having important needs, and neither does is discuss how to solve acknowledged long-term impacts of the environmental problems (externalities) arising from agriculture. It is a document confined within politically acceptable boundaries, failing to emphasise any sense of urgency. Rather, it seems to reassure that existing paradigms can continue without need for investigation and that only tweaks to externalities are needed to ensure continued growth.[19] The report demonstrates the lack of research on integrating local and regional biodiversity into methods of combating pests and land degradation or alternative perennial-based industries. It also assumes that importing resources such as phosphate can continue ad infinitum.

Australian and international members of the Resilience Alliance propose that sustainability is not an endpoint. Ecological systems and human agro-systems should be viewed in terms of their resilience and adaptive abilities. Whether an ecosystem or SES can maintain resilient structure and function when confronted with change-inducing shocks/surprises is related to three parameters of *resilience*: resistance to change or buffering capacity during periods of stress; the ability to return to the original state following disturbance; and the ability to persist despite change drivers through time. Change can also result in complete reorganisation (regime change) into new ecosystem states, humanly undesirable and very resistant to change.[20] For example, huge areas of pastoral lands have lost their topsoil in sheet erosion events due to overstocking. These dysfunctional areas are so resilient to natural plant recolonisation and revegetation that expensive soil disturbance with heavy machinery is necessary to allow capture of windborne organic or silt material. It takes many years to initiate the conditions in which desirable perennial grasses, forbs and shrubs re-establish and help regenerate an A soil horizon.[21]

Some environmental shocks are devastating, and SES cannot adapt. As the ecological systems interwoven with social and cultural fabric disassemble, a feedback loop is created, encouraging further inability to adapt as crucial ecological and cultural relationships break down. The whole now-fragile system collapses into a regime change.[20] As a broad example, Australia's invasion by Europeans and their biological co-invaders introduced devastating shocks to the ancient co-evolved ecosystems and Aboriginal SES. They included new diseases (both human and animal), destruction of food systems and animals that were keystone to co-evolved ecosystems, loss of cultural memory and its critical information, and the imposition of politically driven racial restrictions. So rapid, simultaneous and encompassing were these shocks that Aboriginal SES were unable to adapt. They collapsed, with ecosystem ramifications continuing over the last 200+ years.[22]

Open nutrient and energy cycles

Human food systems can only be sustainable if they are indefinitely resilient – evolving to last for thousands of years or more. In this sense, industrial agro-systems are maladaptive. All food is ultimately derived from plant photosynthesis where energy from the sun, water and CO_2 is converted to sugars that can be used by plants and other life forms that feed upon them. But plants and animals also rely upon soil minerals from Earth itself, using photosynthesised energy to make the components of their cellular bodies then decomposing and recycling this earth matter to make future life.[23] In resilient and sustainable SES, energy and matter are recycled in biodiverse circular systems.

Modern farming systems are based on maximising production with a linear flow system. The matter and energy used to produce and export food and products to distant cities and international markets goes through human bodies or is dumped in landfills, where it is lost as waste into the biosphere far from the farmlands where it originated.[5,24] The economic forces that perpetuate this wasteful approach link farming to financial capital in a manner that favours the money lenders. This can lead to unsustainable outcomes, such as overstocking leading to erosion, to satisfy banking requirements. This addiction to the short-term economic view of industrialised food production was questioned as early as 1909, when Franklin King, retired Professor of Agricultural Physics at the University of Wisconsin, toured China, Japan and Korea to examine how farmers in those countries had produced food for thousands of years without chemical fertiliser inputs. He discovered systems based upon a nearly circular closed-loop nutrient and energy cycle – human, plant and animal wastes were continuously and deliberately returned to farms.[25]

Australian agro-systems provide food for around 61 million people overseas, as well as the local population of about 23 million, so approximately 84 million people in total.[26]

These globalised Australian foods are part of William Catton's 'ghosts', enabling nations to exceed the carrying capacity of their own land by drawing upon 'invisible' carrying capacity and ghost acreages from elsewhere on the planet.[27] Agricultural researcher Murray Fulter agrees, and is of the opinion that agricultural production systems exemplified in the Esperance bioregion lack long-term resilience and sustainability, as they rely significantly upon imported energy and material 'resources that are not managed sustainably and in many cases are pillaging communities throughout the world'.[28] This farmed bioregion is no different from elsewhere in Western Australia, Australia and overseas where imported farming systems use animals and plants adapted to grow in geologically young fertile and trace element-rich soils. These systems are completely dependent upon the continual addition of fertilisers and trace elements. In south-western Australia, essential imported nutrients for animal and plant production include the macronutrients phosphorus, nitrogen and potassium and trace elements such as cobalt, copper and selenium. Cobalt is available locally but is now too expensive to use on pastures. It is mainly exported, as it is a major component in new technologies for green energy. In deficient areas, livestock farmers rely on injecting vitamin B12 (for which cobalt is the precursor) along with selenium three times a year. Plants need the same macronutrients and many other trace elements, including cobalt, for leguminous nitrogen fixation. Other essential inputs are fossil fuels and their products, inputs of lime and the trace minerals along with various pesticide chemicals to control pests and weeds.

For Australia and its agricultural producers, this linear flow of energy and matter by export-dependent agro-systems needs to continually replace the lost nutrients with inputs from 'ghosts' elsewhere, or the systems would fail. The lack of nutrient recycling and reliance on outside (ghost) fertiliser nutrients is a fundamental sustainability problem for Australia.[24]

Attempting to re-engineer ecosystems into functioning in ways that they intrinsically resist, invariably results in unforeseen or unintended consequences causing unfavourable reorganisation.[29] This happened when the Western Australian government released thousands of hectares of land north-west of Esperance in 1978 at Cascades. This was part of its mandated drive to clear more land for wheat and sheep, done without proper soil surveys or land capability assessments and despite researchers warning about salinity. In parts of Cascades, clearing the native ecosystems adapted to harsh soil conditions had disastrous environmental, social and economic consequences. Farmers who took up this CP land found that crop and pasture establishment failed in the first year due to the inherent salinity, alkalinity and sodicity of the soils. They were forced to walk off, and only a few ever received compensation for the money they had spent. The ruined ecosystems were also never 'compensated' with remediation, and the severely degraded land rapidly shifted into a new resilient state where nothing other than a few species of salt-adapted halophytes can grow.[30,31]

When SES are dominated by such maladaptive cultural responses to food production and consumption, their relationships with natural systems will not be resilient or sustainable.[20] The unsubstantiated promises that clearing and re-engineering any land-type will automatically create the conditions for sustainable industrialised styles of agriculture has been so often proven wrong, with profoundly unpleasant environmental and social after-effects in many parts of Australia.

Modern industrial agriculture and externalities

Human economic activities including agriculture are linear, with an extract–produce–use–dump waste and energy flow model.[32] In this model, the environment is a place from which to draw resources, and a sink for pollutive waste – much of which contains nutrients and water that should be returned to their agricultural source. In Australia, already nutrient- and water-poor, they are exported overseas or to nutrient sink cities, never to return. Economists call the environmental and social effects of extractive activities such as agriculture, 'externalities', a human concept based on expediency and false imaginings.[33] Neither producers nor consumers pay directly for agricultural externalities. There are various soulless and theoretical economic arguments that dispute the need to count these unfavourable impacts unless they may lead to market failure, because the market will 'sort it out'. However, in the real world, before market forces can react (they do not behave pre-emptively) to prevent damage, soils and hydrological cycles are degraded, desertification accelerates, wonderful ecosystems and their species are destroyed in the name of economic progress, and animal and human cultures are destroyed along with them. Ultimately, the costs of these interconnected and accumulating impacts will far outweigh the original economic benefits. Climate change is one of the greatest externalities of the Industrial–Technology Age, not only because of the use of fossil fuels but even more due to the rampant clearing, de-greening, hydrological destruction and desertification of the world's landscapes and ecosystems to satisfy the demands of feeding the erupting human population. The concept of externalities should be completely discarded, and all environmental and social consequences included in business production analysis.

Peak phosphate

Phosphorus, potassium, nitrogen and key trace elements comprise 75% of the fertilisers used in Australia. They underpin the ability of all present industrial agro-systems to continue.[24] The world demand for phosphorus is growing at a rate that reflects the need to feed the erupting human population by expanding and intensifying the industrialisation of world agriculture.[34] The remaining easily sourced phosphate rock from which phosphate fertiliser is manufactured will become increasingly scarce and expensive, with varying estimates for when this will occur. Dana Cordell and colleagues predict that the peak of cheap phosphate availability could be reached during the 2030s if human population growth and phosphate demand continues on the present trajectory.[35] Though this threshold seems too distant to warrant concern, it is likely that the ride will be rocky as quality product diminishes. Major producers that supply over 70% of total world phosphate include Morocco (the single largest exporter at 34%), China (the second largest), the US and Russia. These suppliers are likely to manipulate markets as well as tighten supply to meet their own requirements as phosphorus gets scarcer.

Salinity

Dryland salinity is one of the biggest treatable externalities in Western Australia and other parts of Australia.[36] It is currently estimated that over 10% of water is unused in annual cropping and pastures, enough to trigger the pathway to salinisation. Research shows that effective water balance control can only be achieved if over 95% of catchment water in any given year is used. This will only happen if trees, or other perennial plants with similar ecohydrological characteristics as native trees and shrubs, are planted. Without this remediation, millions of hectares of agricultural land in Western Australian and Australian wheatbelts will succumb to salinity.[37]

Soil acidity

In southern Australia, the annual cropping and pasture systems accelerate soil acidification though some native soils such as the sandplain are inherently acidic.[38] Soil acidification significantly lowers growth and production of introduced plants and has been identified by the Western Australian Department of Agriculture as a major concern to future agriculture, with millions of hectares at risk throughout Australia.[19] The target soil pHCa (pH as measured in calcium chloride solution) of 5.5–4.8 (0–30 cm soil profile) is the optimum for crops and pastures. Use of nitrogenous fertilisers on crops and nitrogen fixation by clover and serradella pasture legumes, cause soils to accumulate excess nitrates that cause acidification. Sub-tropical and other perennial pastures can reduce acidification by using nitrogen year-round, but the use of agricultural lime is still considered the best option. This finite resource is becoming more expensive for agricultural use as it is in direct competition with the mining industry.[39]

Chemical farming

Australia has accepted the alarming normalisation of chemical use on grain, vegetable and fruit crops without sufficient research on the long-term health and environmental impacts. For example, the herbicide atrazine has been banned in the European Union since October 2003 because of concerns over its toxicity to water animals and plants, and to human endocrine health including negative impacts on fertility and strong linkages to obesity.[40,41] It is freely used throughout Australia and the US despite ongoing calls for it be banned.[42,43]

Chemical farming has positively ended the need for ploughing, so reducing soil erosion and loss of organic matter. Zero till cropping systems use chemical herbicides instead to kill weeds and animal pests. Fungicide-treated seed is drilled with precision into the undisturbed ground, conserving moisture and soil structure but potentially still impacting soil microorganisms. The normalisation of chemical use means many cropping farmers and horticulturists assert that it is impossible to industrially farm without them, and this is likely true.[44] Target pest species, both animal and plant, are developing strong resistance to pesticides in Australia, requiring stronger chemicals and mechanical technologies to destroy them. For example, rye grass is an important pasture grass – and is the bane of specialist crop farmers because it adapts to any chemical or mechanical destructor used against it. Removing summer water using weeds with herbicides is further damaging regional hydrologies, and in cropping and horticulture the use of high-level toxins, with long half-lives, means they are accumulating in soils and can be implicated in declining soil ecology and linked to numerous human health issues.[45] Chemical use has led to unrealistic requirements of perfection from supermarkets and consumers, a cause of enormous food wastage. Vandana Shiva concludes that multinational chemical businesses such as Bayer, who now own Monsanto, heavily influence the health and viability of farmers, food production, communities and even nations, to their detriment.[8]

Loss of fire control
A feature of the Great Clearing from the 1950s to 1980s in Western Australia was the extinction or near-extinction of many ecosystems, their plants and animals. Aboriginal SES had managed these ecosystems with human physical and mental energy input, with complex fire management and other techniques, to maintain landscape mosaics of ecological communities in multiple states. In colonial times, Aboriginal shepherds continued to use their traditional fire management techniques to regenerate grasses and keep fire potential under control. Early colonists learnt from them how to use fire to manage and maintain grasslands, prevent bush encroachment and mitigate the risk of wildfires. After pastoralism ceased and farming began, with assets such as fencing, fire use was penalised and discontinued.[46] The landscapes became highly wildfire-prone. Mr Campbell describes an unstoppable bush fire in 1913 that started at Dundas (around Norseman) and burnt millions of hectares to the coast at Esperance, over 500 km away.[47] This was previously unheard of. The south-east and south coastal regions, like the rest of Australia, are now plagued by wildfires, with millions of hectares burnt and human and millions of animal lives lost. In 2018–2019, lightning strikes ignited fires on the eastern south coast, burning vast areas including ancient woodland trees with no record of being burnt in the last 200 years. Millions of hectares were burnt throughout western and eastern Australia in those years.[48]

Climate change, desertification and biodiversity loss
There is little understanding or recognition that Australia is undergoing desertification. Evidence exists that before official records began clandestine colonisation in southern and eastern Australia by Europeans and their livestock was already degrading and desertifying native ecosystems, vegetation and soils.[49] Since then desertification has continued in Australia, represented by increasing drought frequency with lowered effective rainfall, triggered by inappropriate land use and poor land management. Sheep numbers in New South Wales rangelands reached 19 million in the 1890s but after the 1901–1902 drought (linked to climate fluctuations and widespread vegetation destruction) they fell to about 3 million. The numbers have never recovered. This pattern was repeated in all states, with livestock numbers declining sharply in rangelands.[50,51]

Desertification is generally associated with arid to semi-arid pastoral rangelands, but it now extends into medium to lower rainfall cleared cropping lands such as Western Australia's eastern wheatbelt. The wheatbelt and broader south-west is undergoing significant rainfall declines of up to 25%. There is sufficient evidence that climate change alone is not the cause, with accumulating impacts from disruption of the hydrological cycle caused by clearing millions of hectares of a once continuous woodland.[52,53]

Biodiversity
Richly biodiverse landscapes have been replaced with fast-growing annual crop species and annual pasture systems of low biodiversity. Todd Dawson and Rae Fry consider that sustainable farms would use systems that closely 'mimic the natural functions of the biota of the region in which the agricultural system is embedded'. This would include using plants (not necessarily indigenous species) that maintain or return hydrological systems to

those that existed pre-clearing.[54,55] Biodiversity comprises a range of plants and animals that maintain landscape-wide soil, habitat and hydrology. Its importance, though making headway overseas, has not entered mainstream agricultural thinking in Australia.[56] Healthy ecoagricultural systems contain important pollinators such as honeybees, native bees, flies, moths, nectar-eating birds and small mammals; smaller pests are controlled by insects, birds and bats; larger predators such as wedgetail eagles and dingoes control larger prey, ranging from small to large vertebrates such as lizards, birds, insects, mice, rabbits, wallabies, kangaroos and medium-sized mammalian predators.[57] Healthy soils contain a vast range of microorganisms and small organisms that help fix carbon in organic matter and supply nutrients to plants. Real sustainability requires an end to the linear flow of nutrients, water and energy, and their return to farmlands. Long-term sustainability with positive ecological relationships is not optional, but a core part of long-term food production systems.[58]

The modern problem of unlinked natural and social capital

Sustainability is a collective concept so production systems must be sustainable over continents, not just on single farms. Farmers must find ways to reduce the damage they cause beyond their own farms, and look at the broader region.[36] But society at large must also be accountable, for its continued existence is dependent upon a sustainable ecoagriculture. Agriculture can no longer be viewed as a collection of discrete independent problems, but as an integrated system impacted and impacting upon local and global socio-economic and eco-environmental factors. This highlights a major difference between past and present SES in Australia. Aboriginal SES were directly integrated and aware of consequences of human–nature interactions for environment impacts were quickly and directly felt, and inappropriate behaviour could jeopardise a group's future. Modern SES illustrate a lack of community-wide awareness of environmental accountability, sharing of problems and adaptive responses, and a disconnection between economic aims, resilience and sustainability.

The Esperance bioregion, like other isolated Australian farming regions, needs physical connections to outside markets for monetary income as the amount of food produced is too great for local consumption. Farm income is made from commodity-based production, with prices based on global pricing regimes.[59] Farming systems and markets are subject to political interference internationally as well as locally, often linked to social perceptions and pressures to change markets and practices from sometimes poorly informed majority city-based voters.

In Australia, a neoliberalist and economic rationalism mindset has put economy and competition before ecological and social sustainability. Frank Vanclay and Stewart Lockie discuss how Australian farmers have had 'market discipline' unilaterally imposed by the economic rationalists whose ideologies include trade deregulation in a globalised world of free markets.[60,61] These strategies have altered the social-ecological nature of rural Australia, causing ongoing hardship. For instance, cost-price squeezing is impelling farmers into continuous cropping systems to increase income, often to buy more land. Agricultural economist Harvey Jones describes how decisions about what should be grown for whole-farm profitability and an economically sustainable return are not based on long-term land capability and sustainability. They are being determined by factors such as commodity markets, cultural acceptability, perceived consumer benefits from competition, distance from markets and global production/overproduction cycles.[62]

This ideology of a world 'free market' fails to account for damaging social and environmental externalities that escape market analysis. It fails to recognise that Australian farms have inherently poorer soils and a uniquely challenging and variable climate. Most other nations heavily subsidise their agriculture, not sharing Australian neoliberal beliefs in the capacity of the free market to regulate their agro-systems. In Europe, it is recognised that food security is not a given and centuries of experience have shown how easily it can be lost during war. In Australia, food producers mostly bear production costs alone, with environmental remediation and sustainability at the bottom of their expenditure list.[59] Since income earned per hectare is the dominant driver of a farm business, only when people are economically comfortable will they think of environmental works.[16]

Recently, some countries have shifted from viewing deregulation as an essential component of globalisation and restructuring. The UK White Paper on International Development, published in 2000, for example, states that 'open trade is not unregulated trade'. A report published by the Organisation for Economic Co-operation and Development (OECD) in 2014 examines agricultural policies in its member countries and the 'producer support estimate' which measures the total subsidies farmers receive from government as a percentage of farm revenue. For Australia, the estimate is 3%, the second lowest support base other than New Zealand, among OECD countries. Compare this to a major competitor, the US, which operates with an average 10% subsidy and has far better soils and climates.[63]

Poorly designed regulated marketing systems invariably have negative impacts. The reserve price scheme for wool encouraged environmental damage as farmers overstocked sheep in conditions of both poor and booming prices.[64] Since deregulation of domestic lamb marketing by the Western Australia Meat Marketing Co-operative in 1994, and of export lambs in 1999, which had cross-subsidised merino lamb producers and homogenised prices, returns for good lambs have achieved their true potential and fostered specialist lamb producers. In contrast, the Western Australia dairy industry performed well for many years under a milk quota system that ensured a year-round stable milk supply for the community, stable milk prices for producers and consumers, and allowed dairy farmers to plan for the long term with at least a middle-class equitable income. Trouble started when milk quotas were allowed to be owned by non-farmers, who then turned them into a tradeable commodity (like water) that the real farmers could not afford. Once deregulated in July 2001 by the Western Australian government in pursuit of the free market, milk producers lost their negotiation power, prices fluctuated frenetically as processors and supermarkets drove producer prices below the cost of production (as has happened elsewhere in Australia) and milk production dropped significantly. By 2017, the number of dairy farmers had decreased by 60%. Farmers left the industry because working 70+ hour weeks with no foreseeable change, with volatile milk prices usually below the cost of production, no power to plan for the future, and an income less than a family on welfare – why stay?[65] So embedded has economic rationalism (market-oriented economic policy) become in political and farmer representative groups that any questioning of its rationale for long-term sustainability is treated as almost heretical.[60] Since it is quite clear that no-one gets paid to be sustainable, long-term food security is unconsidered.

Maximised profitability in the free market system is manifested in agricultural industry research goals. Research paid for by farmers through research organisation such as the Grains Research and Development Corporation, private enterprise chemical/gene technology companies such as Bayer/Monsanto and fertiliser/chemical companies such as CSBP Ltd also follows this mantra, with the added aim of having the organisation's products and services exclusively used by farmers. This has resulted in development of crop varieties that are only productive under high fertilisation levels, technology-led change, genetic modifications to plant genetics that favour particular chemical usage, and inadvertent changes in raw food nutrient levels.

With the emphasis on productivism, there appears to be a lack of research concerning whether the intensively raised crop or animal – pork, poultry, cereal, legume, fruit or vegetable – actually meets human nutritional needs.[66] Emerging evidence indicates that, though apparently well nourished, deficiencies in certain amino acids, fatty acids, vitamins and trace minerals in modern staple diets are leading to forms of malnutrition and increasing rates of chronic disease such as obesity (one of the so-called 'diseases of civilisation'). For instance, modern wheat is measurably lower in micronutrient density than ancient strains; this can be traced to the hybrid dwarf varieties produced as part of the Green Revolution.[67] Changes in wheat protein and carbohydrate composition, along with inappropriate food processing techniques, are being linked to the rise in coeliac disease and non-coeliac wheat sensitivity.[67]

E.N. Anderson describes food demands as resulting from the social and cultural factors that recognise and accept something as food.[68] These are deeply influenced in Australia by the supermarket sector dominated by the Coles and Woolworths duopoly (owned by the Wesfarmers Group and Woolworths Ltd, respectively), which over the past few decades has reduced competition despite the ideology of free market systems. These national retailers have the power to make decisions that not only shape individual consumer choices but also shape food production practices on farms and at retail level, which can affect towns, communities and environments.[i,69] For example, the grain feeding of livestock – chicken, beef, lamb – that started in Australia in the late 1980s is criticised by those who believe the grain should be for human consumption. The development of feedlot enterprises is directly attributable to strict supermarket requirements for uniformity of size, quality and quantities of chicken, beef and lamb with a year-round supply.[70] Feedlotting has benefited grain growers because grain that doesn't meet the strict specifications for human consumption can be sold as feed grain.[71,72] Supermarkets also prefer to deal with fewer large-scale suppliers rather than a multitude of smaller suppliers, reinforcing the 'get big or get out' syndrome.[73] Unlike fruit, vegetables and meat, prices for grain are buffered from the local duopoly because most grain (even feed grain) is exported to countries that are not food self-sufficient. Saudi Arabia, a major importer of Australian wheat, has used up its fossil water that was previously available for grain growing, and now must import wheat for its erupting population.[74]

---

i   In 2007 Coles began to source beef for all its Western Australia stores from Queensland. Western Australian beef producers suddenly lost nearly half of their domestic market. An enormous public outcry and pressure from farmers forced a partial reversal.

Adaptation to the cost price squeeze through land acquisition has increased farm sizes in the Esperance bioregion and elsewhere. Family businesses expand and buy out neighbours, and national and international corporations such as Lawson Grains (Macquarie Bank-backed agricultural fund) and Hassad Australia (Qatar) aggregate farms into large holdings. Chinese corporations now form a group of highly capitalised overseas owners. Land capability can be subsumed by short-term economic drivers forcing unsound environmentally damaging decisions onto the corporate farm managers.[36] For example, a corporate-owned farm in the Esperance bioregion (in Boyatup) was continuously cropped, although its land capability defined it as a pasture-based mixed farm. Despite 2007 being one of the wettest years on record, cropping was rigidly adhered to – and canola germination failed twice. By summer, with no plant cover, thousands of hectares underwent severe wind erosion. The inflexibility of corporate and distant owners can mean an inability to innovate or adapt to environmental and market disturbance.

Esperance Agricultural Officer Kira Tracey describes how farm expansion is resulting in tragic social disruption with declining family-based communities, as small-town populations drop, services contract and local primary schools close. Families have migrated to live in Esperance, and farmers commute to work on their farms.[75] These trends result in a serious loss of emotional land connections. Despite beliefs that larger size produces economies of scale and efficiency, unpublished research by the Department of Agriculture and Food supports overseas research confirming that smaller landholdings are generally more efficient and produce more income per hectare than large properties, especially those that are corporate owned.[76,77]

Until now, industrial agriculture has succeeded, though with uneven distribution, in producing food for erupting human populations but with enormous and continuing costs to Earth's ecosystems. As a system, it is unable to be resilient or sustainable in the long term.

Endnotes

1. Collins P (2014) Ethics of migration. *Sustainable Population Australia Newsletter* 117, September. Collins resigned from the Catholic priesthood over a fundamental discord about the role of humanity on Earth.
2. Berry W (1977) *The Unsettling of America*. Sierra Club Books, San Francisco.
3. Muir C (2014) *The Broken Promise of Agricultural Progress: An Environmental History*. Routledge, London.
4. Saifi B, Drake L (2008) A coevolutionary model for promoting agricultural sustainability. *Ecological Economics* **65**(1), 24–34. doi:10.1016/j.ecolecon.2007.11.008
5. Catton WR (1974) Depending on ghosts. *Humboldt Journal of Social Relations* **2**(1), 45–49.
6. Holling CS, Berkes F, Folke C (1998) Science, sustainability and resource management. In *Linking Social and Ecological Systems*. (Eds F Berkes, C Folke) pp. 342–362. Cambridge University Press, Cambridge.
7. Pretty J (2009) *Agric.-Culture: Reconnecting People, Land and Nature*. Earthscan, London.
8. Shiva V (1991) *The Violence of the Green Revolution*. Third World Network, Penang.
9. Gliessman SR (2007) *Agroecology: The Ecology of Sustainable Food Systems*. 2nd edn. CRC Press, Boca Raton.
10. O'Connor M, Prober S (2010) *A Calendar of Ngadju Seasonal Knowledge: Report 1.2 to the Ngadju People*. CSIRO, Perth.
11. Walker B, Salt D (2006) *Resilience Thinking: Sustaining Ecosystems and People in a Changing World*. Island Press, Washington, DC.
12. Impact Fertilisers (n.d.) *Nutrient Removal by Crops*. <https://impactfertilisers.com.au/wp-content/uploads/impact-calc-nutrient-removal-chart.pdf>
13. Rees B (2019) *Can the Rural Debt and Drought Taskforce Succeed?* Research Report 41. TJ Ryan Foundation, Brisbane.

14. Gro. Harlem Brundtland (1987) *Report of the World Commission on Environment and Development: Our Common Future.* United Nations.
15. Daly H, Cobb J (1989) *For the Common Good: Redirecting the Economy Toward Community, the Environment, and a Sustainable Future.* Beacon Press, Boston.
16. Leadership Council of the Sustainable Development Solutions Network (2015) *Indicators and a Monitoring Framework for the Sustainable Development Goals.* United Nations.
17. Kirschenman F (2010) *Cultivating an Ecological Conscience: Essays from a Farmer Philosopher.* Kentucky University Press, Lexington.
18. Spaiser V, Ranganathan S, Bali Swain R, Sumpter DJT (2017) The sustainable development oxymoron: quantifying and modelling the incompatibility of sustainable development goals. *International Journal of Sustainable Development and World Ecology* **24**(6), 457–470. doi:10.1080/13504509.2016.1235624
19. Department of Agriculture and Food (2013) *Report Card on Sustainable Natural Resource Use in Agriculture.* WA Department of Agriculture and Food, Perth.
20. Davidson-Hunt I, Berkes F (2008) Nature and society through the lens of resilience: towards a human–in–ecosystem perspective. In *Navigating Social-Ecological Systems: Building Resilience for Complexity and Change.* (Eds F Berkes, J Colding, d C Folke) pp. 53–82. Cambridge University Press, Cambridge.
21. Pollack D (2019) *The Wooleen Way: Renewing an Australian Resource.* Scribe Publishing, Melbourne.
22. Scheffer M (2009) *Critical Transitions in Nature and Society.* Princeton University Press, Princeton.
23. Pimentel D (2004) Industrial Agriculture, Energy Flows. In *Encyclopedia of Energy,* **3**. Elsevier, Amsterdam.
24. Brown DA (2003) *Feed or Feedback: Agriculture, Population Dynamics and the State of the Planet.* International Books, Utrecht.
25. King FH (1911) *Farmers of Forty Centuries or Permanent Agriculture in China, Korea and Japan.* Project Gutenberg. Reprinted 2004.
26. Keogh M (2014) Australia exports enough food for 61,536,975 people – give or take a few! *Australian Farm Institute,* 3 September.
27. Catton WR (1974) Depending on ghosts. *Humboldt Journal of Social Relations* **2**(1), 45–49.
28. Fulton M (1993) Cereal and wool production in the Esperance sandplain area of Western Australia: the need for a systems approach for sustainable agriculture. *American Journal of Alternative Agriculture* **8**(2), 85–90. doi:10.1017/S0889189300005038
29. Walker B (2012) Research fellow, CSIRO Sustainable Ecosystems. Pers. comm.
30. Bradby K (2016) CEO of Gondwana Link. Interviewed by Nicole Chalmer, Albany, 26 March.
31. Malcolm CV (1983) *Wheatbelt Salinity: A Review of the Salt Land Problem in South Western Australia.* Technical Bulletin 52. WA Department of Agriculture and Food, Perth.
32. Korhonen J, Honkasalo A, Seppälä J (2018) Circular economy: the concept and its limitations. *Ecological Economics* **143**, 37–46. doi:10.1016/j.ecolecon.2017.06.041
33. Pretty J (2003) *Agric.-Culture: Reconnecting People, Land and Nature.* Earthscan, London.
34. Cordell D (2010) The story of phosphorus: sustainability implications of global phosphorus scarcity for food security. PhD thesis, University of Technology, Sydney and Linkoping University, Sweden.
35. Cordell D, White S, Lindström T (2011) Peak phosphorus: the crunch time for humanity? *Sustainability Review* **2**(2).
36. Simons J (2013–2017) Esperance Regional Hydrologist. Interviewed by Nicole Chalmer, WA Department of Agriculture and Food Office, Esperance.
37. Hatton TJ, Nulsen RA (1999) Towards achieving functional ecosystem mimicry with respect to water cycling in southern Australian agriculture. *Agroforestry Systems* **45**, 203–214. doi:10.1023/A:1006215620243
38. Helyar KR, Cregan PD, Godynet DL (1990) Soil acidity in New South Wales: current pH values and estimates of acidification rates. *Soil Research* **28**(4), 523–537. doi:10.1071/SR9900523
39. Gazey C, Gartner D (2009) *Survey of Western Australian Agricultural Lime Sources.* Bulletin 4670. WA Department of Agriculture and Food, Perth.
40. Bethsass J, Colangelo A (2006) European Union bans atrazine, while the United States negotiates continued use. *International Journal of Occupational and Environmental Health* **12**(3), 260–267. doi:10.1179/oeh.2006.12.3.260
41. Hannink N (2019) 'Safe' herbicide in Australian water affects male fertility. *Pursuit.* University of Melbourne, Melbourne.

42. Center for Biological Diversity (n.d.) *The Case for Banning Atrazine*. <https://www.biologicaldiversity.org/campaigns/pesticides_reduction/atrazine/index.html>
43. Inquiry into Agricultural and Veterinary Chemicals Legislation Amendment Bill (2012) Submission from Friends of the Earth Australia.
44. Harkness W, Harkness P (2016) Farmers at Beaumont. Interviewed by Nicole Chalmers, Beaumont, 20 October. They believe that banning glyphosate would prevent them farming crops in their erosion-prone soil types.
45. Robin M-M (2010) *The World According to Monsanto: Pollution, Politics and Power*. New Press, New York. Many cancers have been linked to pesticide use.
46. Ward D (2010) People, fire, forest and water in Wungong Catchment. PhD thesis, Curtin University, Perth.
47. Campbell D (2006) Scadden. In *Pioneers and Early Settlers of Scadden, 1910–1959*. (Ed. L Bale). Publisher unknown.
48. Department of Environment and Conservation (2009) Fire: the force of life. *Landscope Special Fire Edition*. Vol. 1, December.
49. Gale SJ, Haworth RJ, Cook HE, Williams NJ (2004) Human impact on the natural environment in early colonial Australia. *Archaeology in Oceania* **39**(3), 148–156. doi:10.1002/j.1834-4453.2004.tb00573.x
50. Pickup G (1998) Desertification and climate change: the Australian perspective. *Climate Research* **11**, 51–63. doi:10.3354/cr011051
51. Ludwig JA, Tongway DJ (1995) Desertification in Australia: an eye to grass roots and landscapes. *Environmental Monitoring and Assessment* **37**(1–3), 231–237. doi:10.1007/BF00546891
52. Andrich MA, Imberger J (2013) The effect of land clearing on rainfall and freshwater resources in Western Australia: a multifunctional sustainability analysis. *International Journal of Sustainable Development and World Ecology* **20**, 549–563. doi:10.1080/13504509.2013.850752
53. Kala J, Lyons TJ, Nair US (2011) Numerical simulations of the impacts of land-cover change on cold fronts in south-west Western Australia. *Boundary-Layer Meteorology* **138**, 121–138. doi:10.1007/s10546-010-9547-3
54. Dawson T, Fry R (1998) Agriculture in nature's image. *Trends in Ecology & Evolution* **13**(2), 50–51. doi:10.1016/S0169-5347(97)01251-2
55. Kremen C, Iles A, Bacon C (2012) Diversified farming systems: an agroecological, systems-based alternative to modern industrial agriculture. *Ecology and Society* **17**(4), 44. doi:10.5751/ES-05103-170444
56. Altieri M (1999) The ecological role of biodiversity in agroecosystems. *Agriculture, Ecosystems & Environment* **74**, 19–31. doi:10.1016/S0167-8809(99)00028-6
57. Fischer J, Lindenmayer DB, Manning AD (2006) Biodiversity, ecosystem function, and resilience: ten guiding principles for commodity production landscapes. *Frontiers in Ecology and the Environment* **4**(2), 80–86. doi:10.1890/1540-9295(2006)004[0080:BEFART]2.0.CO;2
58. Williams J, Price R (2010) Impacts of red meat production on biodiversity in Australia: a review and comparison with alternative protein production industries. *Animal Production Science* **50**, 723–747. doi:10.1071/AN09132
59. Clunies-Ross T, Hildyard N (2013) *Politics of Industrial Agriculture*. Routledge, London.
60. Vanclay F (2003) The impacts of deregulation and agricultural restructuring for rural Australia. *Australian Journal of Social Issues* **38**(1), 81–94. doi:10.1002/j.1839-4655.2003.tb01137.x
61. Lockie S (2009) Agricultural biodiversity and neoliberal regimes of agric.-environmental governance in Australia. *Current Sociology* **57**(3), 407–426. doi:10.1177/0011392108101590
62. Jones H (2000) An economic snapshot of the Esperance region. In *Agriculture Western Australia: Esperance Manual*. (Eds P Burgess, M Seymour) pp. 11–13. Agriculture WA, Esperance.
63. OECD (2011) *Agricultural Policy Monitoring and Evaluation 2011: OECD Countries and Emerging Economies*. OECD Publishing, Paris.
64. Smith S (2001) *Deregulation and National Competition Policy and its Effect on Rural and Regional Areas*. Briefing Paper No. 7/01. NSW Parliamentary Library, Sydney.
65. Hale R (2017) Dairy farmers being 'thrown to the wolves'. *Countryman. The West Australian*, 13 April.
66. Sands DC, Morris CE, Dratz EA, Pilgeram A (2009) Elevating optimal human nutrition to a central goal of plant breeding and production of plant-based foods. *Plant Science* **177**(5), 377–389. doi:10.1016/j.plantsci.2009.07.011

67. Shewry PR, Pellny TK, Lovegrove A (2016) Is modern wheat bad for health? *Nature Plants* **2,** 16097. doi:10.1038/nplants.2016.97
68. Anderson EN (2005) *Everyone Eats.* New York University Press, New York.
69. Urban R (2007) Coles' secret multimillion meat deal. *Sydney Morning Herald*, Business section, 26 January 26, p. 1.
70. Keith S (2012) Coles, Woolworths, and the local. *Locale: Pacific Journal of Regional Food Studies* **2**, 47–81.
71. Australian Lot Feeders Association (2017) *About the Australian Feedlot Industry.* Australian Lot Feeders Association, Sydney.
72. CBH Group (2015) *Grain Quality from Western and Southern Australia.* AGIC Singapore, March. The specifications and uses for different classes of grain based mainly on protein content are discussed.
73. Barr N (2011) *The House on the Hill: The Transformation of Australia's Farming Communities.* Halstead Press, Canberra.
74. Nicholas B (2013–2017) Esperance Regional Manager. Interviewed by Nicole Chalmer, WA Department of Agriculture and Food office, Esperance, 10 December 2013–2017.
75. Tracey K (2014) Esperance Development Officer. Interviewed by Nicole Chalmer, WA Department of Agriculture and Food office, Esperance, 10 February.
76. Sackett D (2014) *Corporates can't Replace Family Farms: Sackett.* Farmonline, 10 June. David Sackett is a respected farm business advisor.
77. Lerman Z, Willian RS (2006) *Productivity and Efficiency of Small and Large Farms in Moldova.* Selected paper prepared for presentation at the American Agricultural Economics Association Annual Meeting, Long Beach, 23–26 July.

11
Ecoagriculture for a sustainable future

> Every species [in a forest ecosystem] wants to survive, and each takes from the others what it needs. All are basically ruthless, and the only reason everything doesn't collapse is because there are safeguards against those who demand more than their due … an organism that is too greedy and takes too much without giving anything in return destroys what it needs for life and dies out.[1]

Peter Wohlleben's description of life cycles in European forest ecosystems is an appropriate allegory for the extraordinary ability of human societies to dominate and greedily consume ecosystems without limit, causing their collapse. Throughout human history this scenario has been consistently repeated as populations grew, irretrievably damaged their agricultural systems (causing desertification) and ultimately collapsed.[2,3] Over and over it is implied by writers such as Carolyn Merchant that this disconnect from nature is a recent attribute unique to western society, aligned to the mechanistic 17th-century Cartesian science of French philosopher and mathematician René Descartes (1596–1650).[4] Cartesian science is a reductionist and mechanistic world view where knowledge and wisdom lies in the mathematical analysis of its physical and biological components, and where animals are merely machines and thus incapable of thought or feeling.[5]

Yet it is clear that this human propensity was happening long before modern society, the development of agriculture and western science – all over the world, when humans invaded new lands the ecosystems collapsed, as they did in Pleistocene Australia.[6] Maladaptation has intensified and accelerated with the development of ecologically bereft industrial agriculture and 'civilisation' over the last 10 000 years, and with technological innovation and human population growth in the last century.[7,8] History has shown that the most enduring societies had sustainable ecoagricultural food producing systems interwoven within their social ecological systems (SES). They acted deliberately to keep their topsoils, retain and use natural biodiversity, used closed nutrient cycles, undertook circular economies and limited their populations to below carrying capacity.[2] Wendell Berry, Miguel Altieri and Stephen Gleissman propose that modern societies desperately need to recognise the importance of retaining nutrient cycles and the contribution of biodiverse natural systems to the resilience and continuation of human SES.[9,10]

Recognition of the need for sustainable practices in agriculture can be traced back through ancient Greece, Rome and India. Chinese records of integrated crops, trees, livestock and fish farming date back the furthest, to the Shang-West Zhou dynasties at around 3555–2750 BP.[11] In the early 1900s, F.H. King, Professor of Agriculture at Wisconsin University, travelled through China, Japan and Korea. He determined that intensive agriculture in those societies had been sustainable for at least 4000 years because organic waste, from food to sewage, was returned to the food producing farms to maintain soil fertility. He was convinced that without adopting these measures in the long term, food production in the US was unsustainable:

> … the great movement of cargoes of feeding stuffs and mineral fertilisers to western Europe and to the eastern United States began less than a century ago … These importations are for the time making tolerable the waste of food plant material through our modern systems of sewage disposal and other faulty practices; but the Mongolian races have held all such wastes, urban and rural, and many others we ignore, sacred to agriculture, applying them to their fields.[12]

Another example of a resilient ecoagricultural SES was Incan civilisation before the decimation brought by the Spanish invasion. For thousands of years, hundreds of different plant food species were grown in ways that improved topsoils on the slopes of the steep Andes mountains, using techniques of manuring including guano, highly sophisticated terracing systems that retained rainwater, waste-pit digging, and irrigation systems based on rainfall saved in large cisterns.[13] Building on arable land and the killing of guano-producing sea birds were capital offences. In contrast, since European invasion, the US has lost over 7.5 cm of topsoil from the original 23 cm that its best soils formerly had. Like Australia, its best farmland is being consumed, with unrestrained suburban growth and mining taking precedence over food production.[14]

The disconnection with a new nature is probably implicit when invading new territories or inventing new technologies that mentally, emotionally and physically reinforce distance from nature. This is a consumer front mentality that encourages the notion that there is always an unexploited place somewhere else to move onto.[15] It is not until successful cultural adaptation and co-evolution take place that perceptions of ourselves as part of nature in Australia will become established. The attitudes of some Australian farmers and pastoralists towards their lands are beginning to reflect this as they strive to not only ensure economic sustainability but recognise that achieving this, in the long term, depends upon good land management practices that aim to improve and continuously regenerate soils and landscape biodiversity.

The globalisation of exploitation fosters a consumer front mentality. The resources obtained from 'elsewhere' allow SES that may have collapsed in the past – so relieving their environmental impacts and maybe people even learning from mistakes – to continue with their maladaptive ecosystem interactions.[16] Jules Pretty describes how food is grown in developing countries and exported to Europe, the US and other nations at the expense of impoverished local people and their environments.[17] Importing food and materials and drawing down on water and other resources beyond their capacity for renewal, allows a society to grow far beyond its intrinsic carrying capacity, thus exposing it to violent collapse and war if these become unavailable.[18]

The Syrian civil war which started in 2011 had its drivers developing long before, as maladaptive interactions between people, culture, environment, water and food production reached crisis point. Social, religious and political unrest spiralled as long-term agro-systems collapsed and the economy failed. As Peter Gleick states, 'water and climatic conditions have played a direct role in the deterioration of Syria's economic conditions'.[19] In-depth research looking at the civil war drivers was completed by global strategic analyst Madelaine Lovelle.

She describes how an agricultural collapse occurred because food and water security worsened due to rapid population growth (from 3 million to 22 million between 1950 and 2012). With ground water recharge 30% below the rate of extraction, irrigated food availability dropped by 69%. The 2007–2010 drought forced 1.5 million hungry rural people to converge on cities. Along with inadequate government responses, this provided conditions and opportunities for regime opposition and then extremists to foment conflict.[20]

Duncan Brown proposes that the development of modern agriculture 'brought about a series of ecological changes so profound and so extensive as to [now] place in question the survival of our own species', as the human population explodes and over 75% of planetary ecosystems are undergoing negative human-caused reorganisation.[21] Yet Pretty, Altieri, Bayliss-Smith and others remind us that there have been relatively sustainable agricultural systems.[17,22,23] Many traditional ecoagricultural systems were deeply integrated and co-evolved with the surrounding natural environment, resiliently persisting for long periods until recent industrialised disruptions. Environmental historians Saifi and Drake assert that the agro-ecology of past traditional SES provided food for more than a billion people across the world without causing the local and remote ecological damage of industrialised agro-systems, and without using fossil fuels.[24] This further supports notions that Earth is overpopulated with humanity and that only unsustainable industrialised agricultural systems that are using resources in a non-renewable fashion can support present populations. Even the current agricultural systems, however, cannot support unfettered future population growth.

Since European colonisation of Australia, there has been minimal co-evolution with Australian landscapes. Only the industrialised types of food producing paradigm are recognised as able to support the locally growing population and food exports. The vast investments of physical, cultural and economic infrastructure that support present paradigms are powerful barriers against trying sustainable ecoagricultural systems.[25,26] The sustainability of food, culture and nature are being sacrificed by demands for short-term profit hinged upon consumerism and waste. The rights of other lifeforms to have places to live, eat and breed other than in the margins of reserve lands and national parks, is unacknowledged. As environmental writer David Wallace-Wells comments, there was-is a 'genocidal indifference to the native landscape and those who inhabited it'. This was not unique to Australia, it was a universal element of the European diaspora.[27]

William Catton maintains that the technologies of industrialisation are dangerously reducing the carrying capacity of Earth because of compounding impacts that are linked to globalised exploitation and climate change. These technologies, including new chemical developments, genetically modified organisms, technology and computer science, have created a belief that all limits can be transcended to satisfy expanding resource appetites, and ever-higher standards of living. 'Feeding the world' aspirations do not support researching ways to limit unsustainable population growth. It seems that without deliberate interventions this paradigm will persist for as long as new technological solutions can be found, inputs from Catton's 'ghosts' are not exhausted or until social-cultural and environmental boundaries are reached, causing a violent collapse.[28]

Sustainability as a long-term resilience approach

Johan Rockström and research colleagues have recently developed a new way of looking at the sustainability and resilience of human SES. They propose the concept of 'planetary boundaries' rather than the 'limits to growth' concept which only proposes minimisation of negative externalities resulting from human activities. The planetary boundaries define the biophysical boundaries estimated as the safe space allowable for human activities, before tipping points are reached that would abruptly trigger non-linear uncontrollable major environmental change on a worldwide scale.[29] They have identified at least nine earth system processes which represent the biophysical planetary boundaries:

1. climate change – uncontrolled climate change, as tree and perennial vegetation clearing and desertification accelerate, and industrialisation expands use of fossil fuels;
2. oceanic acidification – as saturation point is reached, billions of tons of CO_2 will be unable to be neutralised by oceanic processes;
3. stratospheric ozone depletion – ozone-destroying human-manufactured compounds;
4. bio-geochemical nitrogen cycle and phosphorus cycle – removing millions of tons/year of nitrogen from the atmosphere, while at the same time polluting oceans and waterways with millions of tons/year of finite phosphorus;
5. global freshwater use – greater than the rate of recharge;
6. land system changes – wide-scale clearing, particularly for growing crops such as cereals, soy and palm oil, other forms of agriculture and urbanisation;
7. biodiversity loss – rate of species extinctions is over 100 times the background rate;
8. atmospheric aerosol loading – from industrial pollution and fire smoke on a regional basis;
9. chemical pollution – consisting of concentrations of toxic substances, plastics, endocrine disruptors, heavy metals and radioactive contamination into the environment.

Seven of the boundaries have had enough scientific research done for understanding and their quantification. The two that, as of 2009, lacked insufficient information, were chemical pollution and atmospheric aerosol loading. Since then information has grown rapidly, including in relation to plastics and atmospheric pollution impacts. Current modelling shows that at least three planetary boundaries have been crossed:

- uncontrolled climate change, as population pressures accelerate clearing for food production, desertification expands and industrialisation grows in developing countries;
- accelerated rates of species extinction, as more and more ecosystems are destroyed for industrialised agriculture, mining and urbanisation;
- the natural global nitrogen cycle, disrupted by desertification and artificial nitrogen production for industrial agriculture, is seriously acidifying soils and polluting oceanic and freshwater systems. Concurrently, finite phosphate is being wasted as pollution of fresh and coastal oceanic waters.

These crossed boundaries could be mitigated if taken seriously; however, with humanity dominating the bulk of world ecosystems and continued population growth, alleviations are unlikely and tipping points may have already been reached.

Before adaptive ecoagriculture throughout Australia can occur, cultural and neoliberal productivism farming and societal SES must change.[30] There is a resurgence in smaller diversified farming systems re-establishing society–food connections through marketing directly to customers at farmers' markets or online. Properly arranged cooperative groups that improve farmer income can stimulate local economies, with social and ecological benefits within communities.[31] Social movements, such as the Slow Food Movement which began in Italy, oppose the degradation of culture and environment exemplified by industrialised fast foods, aiming to raise people's awareness of wholesome and sustainably produced local food.[32] In Australia, organisations such as Sustain: Australian Food Networks, are raising awareness and encouraging urban communities to produce sustainable fresh food at the local level. The Australian Institute of Ecological Agriculture promotes agro-ecology and regenerative agricultural systems to larger-scale farmers as a method to place ecology into farm systems.[22] The independent not for profit group 'Sustainable Population Australia' aims to raise awareness about population growth and the need to determine what population size Australia can comfortably support (https://population.org.au/).

Australia is without a national long-term food security plan based on sustainable and regenerative ecoagriculture. Abilities to change are limited on most farms and pastoral enterprises because of dependence on individual environmental awareness and economic capacity to withstand income loss, whether temporary or longer term, during periods of change.[33,34] Research shows that customers are so used to cheap food that they are unwilling (or sometimes unable) to pay for extra costs passed on from sustainable production practices. Therefore, waiting for market-driven change will be too late.[35] To ensure that sustainability and land stewardship practices become core and integral to food production systems, farmers alone cannot bear the cost. Since communities at large will also benefit, they should also pay. A possibility is to subsidise farmers and pastoralists who switch to recognised sustainable systems, in addition to offering long-term stewardship payments to manage and increase biodiversity.

The federal government attempted a National Food Plan for Australia, released in May 2013.[36] However, it concentrates on food manufacturing, distribution and consumers rather than on the fundamental of food growing systems. There is no recognition of the urgent need for long-term food production planning or what future agricultural landscapes in Australia should look like.[35] At the agro-political scale there appears to be little understanding that regime shifts are already in progress and that long-term planning is overdue.[37]

Long-term transformations of the capitalist economic system to terminate endless growth and unsustainable resource exploitation paradigms would be ideal. For this to happen, there must be admission that these factors are fundamentally responsible for climate change and the other planetary boundary threats. Without such transformations, the ability of individual farmers to adapt could be largely insurmountable and there will be continuing regime shifts into new, less desirable states. Jane Shepherd, professorial fellow at RMIT

University in Melbourne, has re-examined the basis of agricultural systems and neoliberal policy in Australia. She concludes that 'insufficient consideration is given to the unpaid environment, social, community and human-health costs resulting from industrial scale agriculture'.[38] This failing is happening worldwide. For example, Kenya has undergone a change from its long-term closed-loop traditional food systems to export-based systems using bought-in fertiliser and chemicals. Consequently, national and individual food security has worsened.[39] Within Australian Aboriginal SES, everyone had a role in future food planning, being taught from an early age how important it was not to overexploit ecosystems. In contrast, western society by default takes unlimited food for granted, placing its production low on the educational, social and political agenda.

The type of SES in which a culture is embedded indicates its resilience and sustainability. The modern allegiance to human population growth as a vital part of a growth economy, which needs the consumption of more and more 'stuff' by an ever-growing population, is so firmly established that proposals to consider population and growth limits are viewed as against human rights, and even heretical.[40] Along with our modern reliance on 'ghost' economies, they reflect an inherently unstable and unsustainable cultural approach to living. With humans and their livestock now comprising 96% of the mammalian biomass on Earth – only 4% is wild – we are destroying and crowding out other species in a cumulative mass extinction event.[27] Any doubts about this were eliminated with research conducted at the radiation zones of Chernobyl and Fukushima, which have been abandoned by humans. Despite radiation, ecosystems were found to be thriving, and previously absent wildlife were prolifically breeding and re-colonising these landscapes that were free of humans. The study emphatically concluded that it is human presence which consumes ecosystems and takes over habitat, preventing other animal species from living in a particular area.[41]

New approaches
Ecological literacy
Jules Pretty is concerned that society and its food systems display immense ecological illiteracy. This becomes clear when compared to Aboriginal SES and other indigenous food systems whose methods are based upon integration with biodiverse ecosystems.[17] These ecoagricultural traditional systems, with multiple varieties of plant and animal food sources, in which animals are the principal plant harvesters, can offer us landscape-scale solutions – remembering that their purposeful ecological management was for food production. It just so happened that in Australia, what the early European explorers and settlers exclaimed over as beautiful *wild* landscapes rich in animals and plants were in fact Aboriginal SES that illustrated a most effective way of ensuring long-term food sustainability and resilience.[42,i]

Ecological literacy is desperately needed in our society and should be an inevitable part of childhood education and all varieties of tertiary education. Rather than the present focus on productivism, ecological literacy should underpin all agricultural education. It is essential for all sectors of society to understand this concept. Ecological literacy is intimately involved

---

i Aboriginal peoples view the landscape as an integrated source of food production.

with the cognitive processes of adaptive learning, continuously shaping, knowing and adapting to our environment to form a relationship as part of nature.[17] Ecoagriculture involves learning to read the intimate nuances of local environments and how to adapt to them, using constant empathetic observation and appropriate documented research findings that fit within sustainable outcomes (such as optimising soil life, use of lime, gypsum and fertiliser, and grazing or cropping systems that incorporate wildlife habitats).

Perennialisation

Regime shifts such as the hydrological changes that cause salinity and desertification can be solved if there is resolution to return over 90% of Australian human-impacted landscapes to perennial plants, not all of which have to be trees. This could initiate exciting new farming systems based on livestock grazing perennial grasses and shrubs, perennial crops and salt-tolerant perennials. Though this is happening on a small scale in Western Australia it is usually part of annual cropping with salt-tolerant perennials on the salinising water courses.[43] Many of the early planted salt-tolerant species funded through Landcare are now dying from hypersalinisation. Not enough of the cropping landscape was planted with perennials to prevent the rising water table.

According to forest ecologist David Ellison and colleagues, perennialising the landscape (especially with trees) should be the primary concern of all decision-makers. There are vast climatic positives that come through cooling local and regional climates and increasing bioprecipitation (biologically mediated rainfall), with carbon sequestration an automatic side benefit.[44] To provide incentives to perennialise the landscape, direct payments to landholders through a worthwhile price on sequestered carbon should become mandatory. This approach would need to be flexible as to types of land use that can do this, including not only tree reforestation but also carbon farming in which livestock graze perennial grasslands and dual-purpose perennials, such as *Microlaena*, that produce grain as well as fodder.[45,46]

Arable farming will in the long term need to convert from annual to perennial cropping systems that create healthy carbon- and microbial-rich, water-absorbing soils.[47] As plant researcher John Pickett explains, the perennialisation of crop species should be at the forefront of agricultural research. Despite the anti-genetic modification sentiments expressed by a large section of the community, plant bioengineering may be needed to achieve this, as most crop species are derived from annual grasses and legumes.[48] Alternatives include breeding for greater grain yield in existing perennial plants and accepting them as legitimate foods at a societal level.

Agricultural systems are cultural ecosystems

Pastoralists and farmers who embrace ecoagriculture with regeneration and sustainability in mind seem to have some common attributes. Often, they've had an awakening experience of deep stress, precipitated by a mix of financial and family problems and severe land degradation, leading them to question everything they have been doing. Previous peer group pressure to accept the normative is discarded, opening them to other possibilities. Financial success is no longer the main driver; instead, it changes to an ecologically based striving to achieve healthy soils and landscapes. The journey of turning stations and farms into self-regenerating ecoagricultural systems will inevitably become profitable. The problem is how to survive economically until regeneration starts to function.

Rangelands

Australia's pastoral rangelands comprise those areas where rainfall is too low, variable and unreliable for more intensive farming systems. They include nearly 80% of all Australian lands as savannahs, woodlands, shrublands, grasslands and wetlands.[49] Most are overwhelmingly degraded due to persistent colonial mythologies about how to wrest income from them with livestock.

Rangelands in Western Australia comprise the Southern rangelands, Pilbara and Kimberley pastoral districts. Long-term sustainable production requires maintenance and improvement of soils using perennial and annual plant cover ranging from native grasses, shrubs and trees, to useful and beneficial introduced species. The greatest tragedy of degraded rangelands is ongoing desertification, with Australia-wide climatic impacts. Drought is very often a human concept exacerbated by and resulting from human activities (such as overstocking) that have disrupted the ability for soil water absorption and retention. There are faulty government and individual responses to dry conditions, such as handfeeding to retain livestock that then further degrade the landscape, and a lack of understanding that 'average' or 'normal' seasons are not guaranteed with Australia's variable climates.

I spoke with several successful land managers who intuitively display high levels of ecological literacy – and they showed that pastoral landscapes can be managed relatively sustainably and produce income.

Kim Parsons, from Coolawanyah Station near Tom Price, explained that observing interactions between grass species, climatic conditions and cattle management is necessary for maintaining healthy pastoral landscapes.[50] He admits to lessons learnt when grazing sheep, and has had greater ongoing success since changing to cattle and adopting a more land-friendly approach, intuitively based upon a lifetime of learning from past ecosystem observations of what has or has not worked in similar conditions. He describes native grasses that grow in better soils, including Roebourne Plains grass (*Eragrostis xerophila*) and Mitchell grasses (*Astrebla* species) as good cattle feed, though less resistant to grazing pressure than the introduced buffel grass (*Cenchrus ciliaris*). Buffel grass, which is native to Africa, the Middle East and Asia, was introduced in the 1870s by Afghan cameleers who used it to stuff their harness and packs.[51] Pre-adapted to grazing by large ruminant herbivores, buffel grass has expanded grazing areas as it will grow in poorer areas such as soft spinifex (*Triodia* species) country and pindan soils. If pastures are managed for the less resilient but better-quality native grasses, buffel grass has an important role in the pasture mix. Like other approaches (including those of indigenous societies) that have longer-term success, this approach is conservative, aiming at optimisation rather than maximisation. Kim stocks 25% below the Agricultural Department's suggested carrying capacity (based on an effective summer rainfall and the lease being fully developed for grazing) even with other large herbivores (kangaroos, horses, camels) under control. He describes the departmental recommendations as based on average seasons (whatever they are!) with flexibility depending on seasonal variations. The danger is that adhering to them can quickly turn to disaster in poor years as the recommended levels then become 'overstocking', causing rangelands to become denuded and damaged.

David and Frances Pollock, on Wooleen Station near Meekatharra, are even more conservative in their expectations of country. They destocked cattle and sheep and allowed dingoes to return. This effectively controls overgrazing pressure from kangaroos, eliminates wild goats and helps rangeland regeneration. Dingo predation has significantly reduced soil erosion and improved water retention in the landscape, allowing perennial native grasses to spread, the re-establishment of palatable shrubs and increased survival rates of tree seedlings.[52] More water in the landscape allows a strong resilient hydrological cycle to recommence involving plants, soil structure, fertility and organic matter. On Wooleen, there are water-slowing enviro-rolls hundreds of metres long, sited across water flow paths. After large rainfall events, these offer sediment/seed drop points that are good places for plants to re-establish, as well as hydrating the wider landscape. Store cattle are now being opportunistically grown out on regenerated grassland areas, but permanent cattle breeding is no longer part of their operation.

Bob Purvis describes his father's two reasons for taking up Woodgreen Station north-east of Alice Springs. First, as it had no surface water it was not permanently inhabited by local Aboriginal people and so he hoped his ownership was unlikely to cause problems between himself and them. Second, it was particularly good country for horse breeding.[53] He ran merino sheep for wool and bred horses for the Indian remount market. However, in ignorance he overstocked the property, degrading the best country so that the soil A horizon was swept away by wind and water erosion, leaving the soil B horizon which is inhospitable for plant re-establishment. Bob has spent the last 50 years regenerating the deserts created during his childhood. He has developed a relatively economic system to recover the soil A horizon by creating soil-correcting banks in areas that had been badly degraded. Using a dumpy level to work out landscape levels, and his bulldozer consistently one day a week, he rips along the levels in the hardened pavements of the soil B horizon and uses the loosened soil to build banks 100–150 m in length. The banks can hold up to 23 cm of water and trap wind- and water-borne silt, seed and fine organic matter. After five to eight years there is enough material to allow annual grass establishment, and eventually the thickening soil humus layer lets perennial grasses follow. The plants are then able to continue the soil-building process by trapping silt, and their root exudates encourage soil biology. Over time, a new A horizon develops.[54,55]

Bob believes the Northern Territory Pastoral Lands Board had unrealistic, economic-based expectations of productivity rather than a focus on land capability, and considers it largely responsible for the earlier desertification of Woodgreen and other stations in the region. The Board's recommendations were originally based on brief observations rather than long-term trials. In Bob's opinion, they should be revisited to consider true land and managerial capability and redress the tendency of bureaucrats and land managers to overestimate the productiveness of this ancient land.[53]

Woodgreen today is a good example of how to regenerate and manage these types of rangelands. There are 29 paddocks, each designed to include good country with at least 10 palatable species. At any one time 11 paddocks are in use by the cattle who are divided into three mobs. Each mob has two to three paddocks to rotate around, giving paddocks rest periods of up to a year. Slowly the grasses, shrubs and trees have returned. Fire is used annually to cool-burn patches and reduce risk of very hot destructive wildfires after exceptional growth years. These patchy burns have recovered grasslands and allowed growth of important tree species, including the nutrient-recycling bloodwood trees (*Corymbia opaca*) over the whole property where unpalatable bush once dominated. Dense *Acacia aneura* thickets growing in the best limestone country, once controlled by Aboriginal fire use and burrowing bettongs, are being opened to perennial grasses again. The introduced buffel grasses – US and Gaynor – have significant roles in covering bare poorer country, preventing wind erosion, regenerating soils and improving the grazing mix and stock carrying capacity.[53]

Bob observes that the only way to run arid to semi-arid lands is to learn the ecology and soil preferences of the native grasses and to manage livestock and wildlife accordingly, rotating when necessary and always looking after the land's ecosystem processes *before* expecting monetary returns. It is always better to understock as overstocking makes less money per beast, degrades the landscape and compounds the problems.[55]

These case studies on severely damaged lands show the long-term difficulties involved in regenerating them, and how much sacrifice unfunded individuals have been making to do this. Under the present governance, they are often fighting bureaucratic and political systems which have inadequate ecological knowledge or any broad perspective about the urgency to do this everywhere in Australia. Healthy productive rangelands are the borderline defence against creeping desertification and national rainfall declines exacerbated by climate change. Large-scale well-funded rangeland regeneration, under proven management systems, is vital for intensely softening local and regional climate change impacts, improving habitats for biodiversity and providing huge potential for carbon sequestration.

Agricultural farmlands

All intensive food production in Australia take place upon ancient strongly weathered soils, mostly infertile compared to the young soils of Europe and the Americas where most agricultural plants evolved. Though soils worldwide hold the largest store of organic carbon in terrestrial ecosystems, natural sandplain soils of southern Western Australia have low soil organic matter (SOM) and can be poorly structured, with carbon mostly retained in the native plants. Agricultural and pastoral activities have further destroyed structure and function.[56] Generally, eastern Australian soils are younger and more fertile due to more recent volcanic and tectonic activity.[57] Improving native soils for introduced plants is achievable by pursuing activities and practices that build SOM with soil carbon as both organic fractions and phytoliths.

Grazing cattle and sheep on perennial pastures can improve biodiversity and soil much better than specialist large-scale cropping and horticultural systems. Perennial pastures can sequester enough carbon in SOM and plant phytoliths (PhytoC) to negate greenhouse gases produced by the herbivores grazing them.[58,59,ii] Cattle are regularly accused of 'destroying the planet' but perennial grazing systems that integrate complex natural and revegetated mosaics, have an integral part in soil regeneration and reversing climate change.[60] Biodiverse pastures contain many cohort species such as serradellas, naturalised and improved clovers and many edible grasses, both annual and perennial, as well as broad-leafed plants (weeds in cropping systems). Along with SOM and carbon sequestration, soil fertility is enhanced from clovers fixing atmospheric nitrogen, organic matter breakdown and the dung beetle activity of burying animal manures into the deeper soil profile. Perennial pastures increase soil water infiltration and, with their mulching effects, retain soil moisture for greater biodiversity in soil microorganisms, mycorrhizal fungi, beneficial insects and mixed tree–grassland ecosystem birds, small and larger mammals.[61,62,63] On Coronet Hill we no longer use pesticides to attempt elimination of red-legged earth mites (*Halotydeus destructor*) – which are now highly resistant in Western Australia due to pesticide overuse. The non-use of pesticide has allowed the proliferation of predatory mites, lacewings, beetles, spiders and ants that, along with grazing management, control red-legged earth mite numbers below damaging levels.[34]

On the Esperance sandplains, research has found that (other than applying phosphorus) the most positive transforming process is to apply 200–400 tonnes/ha of suitable clay to sandy soils. This permanently changes their water- and nutrient-holding capacity and the soil structure, increasing soil biology and the ability to build carbon with SOM.[64] On Coronet Hill, clayed sandy paddocks have doubled in cattle carrying capacity, and as long as regenerative practices and fertilisation are maintained these changed soils become highly productive bases for introduced agro-systems. There is potentially further production capacity if the right mycorrhizal fungi are present. Overseas research indicates that particular taxa are plant species-specific and can significantly increase plant production.[65] These interactions, as well as the roles of native fungi and mycorrhizal fungi in ecoagriculture, appear to be a neglected research area in Australia, where so much research effort is instead devoted to chemical and technological methodologies for productivism.

Activities such as ploughing and leaving the soil surface uncovered over summer are physically and biologically damaging to soils. Bared soil can heat up to over 60°C, killing surface soil organisms or significantly reducing their activities. Pesticide chemicals can kill beneficial soil and predatory organisms that inhibit or prey upon root disease and other plant predators. Their yearly application is seemingly done without reference to the half-life of previous applications, which can be up to 50 years.[66,67] Minimal tillage cropping systems that include stubble retention and cover cropping can use more water, improve microorganism survival, reduce soil compaction, eliminate soil erosion and help towards healthy hydrology in the landscape.[30]

Wendy and Peter Bradshaw of Murray Wells farm near Tambellup, north of Albany, are farming for landscape and soil biodiversity and less reliance on artificial fertiliser and chemical inputs. Wendy's descriptions of their efforts towards sustainable food production on their sheep and crop farm, reflect a deep connection with the land. Perennial plant

---

ii   Plant phytoliths (up to 3% of total soil mass) are highly resistant to decomposition compared to other soil organic carbon components. They can last for upwards of 1000 years as a stable compound of silica and occluded carbon.

biodiversity has been increased by revegetating about 100 ha along fencelines and natural water courses with mixed plantings of native trees, introduced perennials and understorey. This type of revegetation, rather than the typical monoculture tree planting, is the most useful to nature. It provides year-round habitat for the native pollinators and predators which are so important for crop and pasture production, and minimises pesticide use. Wendy and Peter are also striving to increase their SOM by rotating crops every three to four years with merino sheep pastures but have found it very difficult to grow agricultural plant species without artificial fertilisers, especially phosphorus.

Karingal farm, north of Esperance, specialises in cropping. The Campbell family have been experimenting there with alternative cropping methodologies for 24 years. Their journey was triggered when the mother, Linda, was diagnosed with non-Hodgkin's lymphoma, a cancer of the lymphatic system. It is often called the 'farmers' disease' as farming has been shown to be among the highest risk factors. Research is gradually determining the pesticides that may be causally linked to this cancer.[68]

David Campbell had noticed that the native mallee soils had far more visible fungi than his cropped soils, which he attributed to use of highly toxic chemicals and loss of organic matter. This helped in changing his views about 'good' farm practice and he decided to go against the industrial cropping norm, using softer chemicals and finding ways to improve SOM, soil biology and grain nutrient density. Since returning to the farm, sons Brad and Greg have continued exploring this pathway. High-quality compost was made for six years and was found to help improve grain quality. However, after the compost was destroyed in an enormous district wildfire the economics were assessed and they concluded that grain prices were not high enough to justify the expense of making and applying broad-acre compost. They are now concentrating on practices that have worked well for other regenerative farmers. Minimum tillage is long accepted, but this does not continuously improve soils because, unless combined with deliberate efforts to improve soil biology, a SOM limit of around 1.2–1.5% is reached in Esperance soils.[64] The Campbells are aiming for higher SOM levels by keeping plant cover year-round on their soils with heavy stubble, reducing the need for pre-emergent weed spraying, and using legume-based crop rotations and multispecies cover crops including legumes and grasses. They consistently test for soil biology and minerals, adding minerals when necessary, and use simple forms of fertiliser such as superphosphate rather more processed forms. They have become convinced that, without applying phosphorus to the naturally deficient soils, farming in this region is realistically impossible. The improvements in Karingal farm's soil biology are so great that David says, 'my livestock are called earthworms'.

Ideas for new diversified perennial food systems

To have some control over a likely collapse and reorganisation of industrial agricultural SES in Australia, we must make dramatic and fundamental cultural changes both locally and globally. In Australia, large areas need to be regenerated with a range of perennial plants producing quite different long-term foods and other resources. Agricultural landscapes need to be redeveloped in a diversified manner that promotes ecosystem biodiversity with intensive food production mosaics, interspersed with less intensive systems.[30] Animals that are presently disdained as 'ferals' and 'pests', including camels, horses, donkeys, red and grey kangaroos, euros, wallabies and rabbits are intrinsically adapted to lower nutrient levels and need less trace elements than traditional domesticates. They, along with cattle, can also withstand dingoes as controlled apex predators which would allow the ecologically important wild to contribute as part of human food production. Such systems could prove more sustainable as new sources of meat, milk and fibre. On Coronet Hill, sustainably harvesting over 500 kangaroos a year for human consumption would be completely feasible if we were paid enough for each animal, and local mobile or on-farm processing systems allowed.

Future grazing could be based on drought- and heat-resistant perennial pastures, shrubs and trees. If all nutrients taken off farms are returned, intensive agriculture close to towns and cities may be possible, as proved by the intensive nutrient recycling of traditional Chinese food systems.[12] Unless nutrients can be returned locally and from overseas, farming as an Australian export industry will eventually be unsustainable.[12]

Australia's native grasses, legumes and shrubs have had millions of years to adapt to Australian conditions of unpredictable rainfall and lower-nutrient soils, so under new climatic and low fertiliser conditions they are likely to be superior to imported species.[69] Native or bio-engineered perennial tree legumes and acacia species have potential for cropping. The seeds of about 50 species of dryland acacia species were a significant seasonal food for Aboriginal people.[70] Research in Africa has shown many of them are easily cultivatable and give heavy yields of palatable protein-rich nutritious seed for human and animal food.[71] Ian Chivers and colleagues have spent years studying and promoting the potential of native perennial grasses as alternative grains.[46] Grasses such as weeping grass (*Microlaena stipoides*), high in protein relative to rice, uses water year-round, and could help prevent and reverse the salinity spectre as a perennial pasture grass that also produces high-quality grain under lower-fertility spectrums. Weeping grass has protein levels of 22% compared to the best wheats at 12–15%, that need copious fertiliser and chemical sprays.[72]

Ecoagriculture consisting of permanent mosaic patches of tree and shrub crops, including rotational woodlots, wattles, drought-tolerant native and introduced nut and food tree species such as pistachio, moringa (*Moringa oleifera*), quandong, sandalwood, timber trees and olives can be slow to bring returns but will last for centuries. With present short-term economics, the question of how to gain a return from the land in the tree phase needs multidisciplinary research; in other places in the world, 50–100 year timber and food tree rotations are accepted as normal and even longer rotations are being considered for carbon sequestration offset payments and species habitat.[73,74] Trees provide enormous ecoagricultural benefits as they increase rainfall (bioprecipitation), extract deep nutrients and return them to the soil surface, dramatically increase ground water use, sequester carbon and provide habitat for useful organisms. The right species should be accepted as an essential part of farm systems in Australia, not as a hindrance. In New Guinea, nitrogen-fixing casuarina species are grown as woodlots then cleared for food growing and replanted, as an essential part of long-term horticultural rotations in use for thousands of years.[75] Eric Toensmeier outlines how Australian farming systems lag far behind tree–pasture–horticulture–tree rotation systems that have been practised for centuries in traditional French and African ecoagricultural systems.[76] Traditional olive production is another model that now has government support, as an alternative to monocultures in Spain.[77]

Rob Sudmeyer and colleagues describe results from phase farming research with eucalypts which indicated significant carbon sequestration and a dramatic lowering of the water tables for up to 18 years after returning to conventional agriculture. Limiting factors included acidification and nutrient loss in the surface soils and the problem of how to make an economic return from the land in its tree phase.[78] Sudmeyer then conducted a small-scale trial at Esperance Downs Research Station using the nitrogen-fixing deep-rooted leguminous tree *Acacia saligna* instead of eucalypts. His results were similar to those in other places where leguminous trees are used – the soil structure improved, carbon was sequestered, nitrogen and potassium were returned to the soil and the water table was significantly lowered. Some farmlands regenerated after removal of Tasmanian bluegum (*Eucalyptus globulus*) plantations have produced at least double the average crop yields. This is probably due to the trash left (leaves, twigs and bark) containing up to 1000 kg/ha of phosphorus.[iii]

In Victoria, salinisation has been successfully reversed using a combined perennial pasture system based on lucerne, along with tree revegetation, which reduced recharge to almost zero.[79] In Western Australia, salinisation of the originally fresh Denmark and Kent Rivers was caused by widespread clearing of native trees and bush in their upper catchments. These were identified in 1996 as potable water recovery catchments under the state's Salinity Action Plan. To reduce salinity, the Denmark River catchment underwent major revegetation and high-water use farming systems (planting of perennial-based pastures) and after 20 years the river is now classified as fresh. Unfortunately, this river's health is not assured, as it depends upon perpetually retaining privately owned tree plantations, subject to economic cycles and stressors that could lead to plantation clearing. Obviously, the plantations should be purchased by government and left in trees in perpetuity.[80]

The factors preventing adoption of such sustainable integrated systems are both cultural and economic. Accepting that present industrialised systems are not automatically superior to proven past systems is a hard belief to break for they do earn immediate farm and export income, especially if the future is discounted. Cultural beliefs about 'what is food' in Australia are expanding but the rate of change needs to accelerate so that 'new' (actually ancient) ecologically embedded foods are recognised as edible and attempts are made to produce them profitably. Jared Diamond has proposed that western society can choose 'to fail' by maintaining the European style agro-systems that have worked in the past, as the Norse did in Greenland despite the climate becoming too cold for those agro-systems. Or choose to survive, by making fundamental adaptive changes to the core values. The people of Tikopia made the latter choice, deliberately controlling their population size and removing pigs from their island (and out of their culture) to prevent overexploitation of limited resources.[2] Early Aboriginal SES also chose to survive, moving from their early extractivist resource culture to one of co-evolved ecoagricultural food production.

---

iii Robert Sudmeyer, Researcher, Department of Primary Industries and Regional Development, February 2018, pers. comm.

Survival is entirely possible but as a society we must come to terms with and admit that erupting human populations and the endless economic growth paradigm are ruining our planet. Environmental stewardship must encompass all of society, not just direct land managers, to encourage and grow synergies between farmlands, cities and non-farmed lands.[7] Agriculture cannot be separated from social systems, adaptability in agriculture must be matched by adaptability in society at large before sustainability and resilience become achievable.

The maladaptive pathways taken since colonial invasion of Australia are mostly looked back upon with pride as European settlers battling against the odds to subdue, change and tame the 'wild' nature that was in fact managed by Aboriginal peoples. In the short term there are many examples of how successful this approach was, and the industrialised agro-systems that developed are considered pinnacles of human endeavour – but only if the externalised costs they continue to incur are not counted.[16] In the final analysis, nurturing healthy viable Earth systems with people requires the diversity of nature as part of food systems. Without this, the liveable and food producing areas in Australia will contract as desertification and sea levels rise, climate change takes hold and societies implode. The only option left for survivors may be to become Aboriginal hunter-gatherers and ecofarmers again.[81]

Endnotes

1. Wohlleben P (2015) *The Hidden Life of Trees: What They Feel, How They Communicate*. Black Inc., Melbourne.
2. Diamond J (2005) *Collapse: How Societies Choose to Fail or Succeed*. Penguin, Melbourne.
3. Flannery T (1994) *The Future Eaters: An Ecological History of Australasian Lands and People*. Reed Books, Sydney.
4. Merchant C (2003) *Reinventing Eden: The Fate of Nature in Western Culture*. Routledge, New York.
5. Barseghyan H, Overgaard N, Rupik G (2018) *Introduction to History and Philosophy of Science*. Pressbooks [online].
6. Miller GH, Fogel ML, Magee JW, Gagan MK, Clarke SJ, Johnson BJ (2005) Ecosystem collapse in Pleistocene Australia and the human role in megafaunal extinction. *Science* **309**, 287–290. doi:10.1126/science.1111288
7. Kirschenman F (2010) *Cultivating an Ecological Conscience: Essays from a Farmer Philosopher*. Kentucky University Press, Lexington.
8. Ponting C (2007) *A New Green History of the World: The Environment and the Collapse of Great Civilizations*. Random House, New York.
9. Altieri M (1999) The ecological role of biodiversity in agroecosystems. *Agriculture, Ecosystems & Environment* **74**, 19–31. doi:10.1016/S0167-8809(99)00028-6
10. Gliessman SR (2007) *Agroecology: The Ecology of Sustainable Food Systems*. 2nd edn. CRC Press, Boca Raton.
11. Pretty J, Bharucha ZP (2014) Sustainable intensification in agricultural systems. *Annals of Botany* **114**, 1571–1596. doi:10.1093/aob/mcu205
12. King FH (1911) *Farmers of Forty Centuries or Permanent Agriculture in China, Korea and Japan*. Project Gutenberg, 2004.
13. Kendall A (1997) Traditional technology emphasized in a model for Andean rural development. *Journal of International Development* **9**(5), 739–752. doi:10.1002/(SICI)1099-1328(199707)9:5<739::AID-JID481>3.0.CO;2-I
14. Goran M (1981) *Conquest of Pollution*. Environmental Design and Research Centre, Newton.
15. Silliman BR, McCoy MW, Angelini C, Holt RD, Griffin JN, van de Koppel J (2013) Consumer fronts, global change, and runaway collapse in ecosystems. *Annual Review of Ecology Evolution and Systematics* **44**, 503–538. doi:10.1146/annurev-ecolsys-110512-135753
16. Catton WR (1974) Depending on ghosts. *Humboldt Journal of Social Relations* **2**(1), 45–49.
17. Pretty J (2002) *Agri-culture: Reconnecting People, Land and Nature*. Earthscan, London.

18. Catton WR (1982) *Overshoot: The Ecological Basis of Revolutionary Change.* University of Illinois Press, Champaign.
19. Gleick PH (2014) Water, drought, climate change, and conflict in Syria. *Weather, Climate, and Society* **6**(3), 331–340. doi:10.1175/WCAS-D-13-00059.1
20. Lovelle M (2016) *The Future Beyond Conflict: Food and Water Security in Syria.* Future Directions International, Perth.
21. Brown DA (2003) *Feed or Feedback: Agriculture, Population Dynamics and the State of the Planet.* International Books, Utrecht.
22. Altieri MA (1995) *Agroecology: The Science of Sustainable agriculture.* Westview Press, Boulder.
23. Bayliss-Smith TP (1982) *The Ecology of Agricultural Systems.* Cambridge University Press, Cambridge.
24. Saifi B, Drake L (2008) A coevolutionary model for promoting agricultural sustainability. *Ecological Economics* **65**(1), 24–34. doi:10.1016/j.ecolecon.2007.11.008
25. Gould SJ (1990) *Wonderful life: The Burgess Shale and the Nature of History.* W.W. Norton, New York.
26. Massy C (2017) *Call of the Reed Warbler: A New Agriculture – A New Earth.* Penguin, Melbourne.
27. Wallace-Wells D (2019) *The Uninhabitable Earth: A Story of the Future.* Penguin Random House, New York.
28. Catton WR (n.d.) *Humanity's Future Imperilled by Cultural Lags.*
29. Rockström J, Steffen W, Noone K, Persson A, Chapin FS, Lambi E *et al.* (2009) Planetary boundaries: exploring the safe operating space for humanity. *Ecology and Society* **14**(2), 32. doi:10.5751/ES-03180-140232
30. Kremen C, Iles A, Bacon C (2012) Diversified farming systems: an agroecological, systems-based alternative to modern industrial agriculture. *Ecology and Society* **17**(4), 44. doi:10.5751/ES-05103-170444
31. Fielke SJ, Bardsley DK (2013) South Australian farmers' markets: tools for enhancing the multifunctionality of Australian agriculture. *GeoJournal* **78**, 759–776. doi:10.1007/s10708-012-9464-8
32. Schneider S (2008) Good, clean, fair: the rhetoric of the slow food movement. *College English* **70**(4), 384–402.
33. McKenzie FC, Williams J (2015) Sustainable food production: constraints, challenges and choices by 2050. *Food Security* **7**(2), 221–233. doi:10.1007/s12571-015-0441-1
34. Williams JE, Price RJ (2010) Impacts of red meat production on biodiversity in Australia: a review and comparison with alternative protein production industries. *Animal Production Science* **50**(8), 723–747. doi:10.1071/AN09132
35. Bhaskaran S, Polonsky M, Cary J, Fernandez S (2006) Environmentally sustainable food production and marketing: opportunity or hype? *British Food Journal* **108**(8), 677–690. doi:10.1108/00070700610682355
36. Australian Government (n.d.) *Towards a National Food Plan for Australia: A Summary of the Green Paper.* Contains large amounts of dated data and stale contact details.
37. Department of Agriculture and Food (2013) *Report Card on Sustainable Natural Resource Use in Agriculture.* WA Department of Agriculture and Food, Perth.
38. Shepherd J (2011) Global food insecurity: rethinking agricultural and rural development paradigm and policy. In *The Self-reliant Country: Sustainable Agricultural Policy for Australia.* (Eds M Behnassi, S Draggan, S Yaya). Springer, Dordrecht.
39. Gitau T, Gitau M, Waltner-Toews D (2009) *Integrated Assessment of Health and Sustainability of Agroecosystems.* CRC Press, Boca Raton.
40. Penn DJ (2003) The evolutionary roots of our environmental problems: toward a Darwinian ecology. *Quarterly Review of Biology* **78**(3), 275–301. doi:10.1086/377051
41. Webster SC, Byrne ME, Lance SL, Love CN, Hinton YG, Shamovich D, Beasley JC (2016) Where the wild things are: influence of radiation on the distribution of four mammalian species within the Chernobyl Exclusion Zone. *Frontiers in Ecology and the Environment* **14**(4), 185–190. doi:10.1002/fee.1227
42. Rose DB (1996) *Nourishing Terrains: Australian Aboriginal Views of Landscape and Wilderness.* Australian Heritage Commission, Canberra.
43. Masters D, Edwards N, Sillence M, Avery A, Revell D, Friend M *et al.* (2006) The role of livestock in the management of dryland salinity. *Australian Journal of Experimental Agriculture* **46**(7), 733–741. doi:10.1071/EA06017
44. Ellison D, Morris CE, Locatelli B, Sheil D, Cohen J, Murdiyarso D *et al.* (2017) Trees, forests and water: cool insights for a hot world. *Global Environmental Change* **43**, 51–61.
45. Helgadóttir A, Hopkins A (Eds) (2013) *The Role of Grasslands in a Green Future: Threats and Perspectives in Less Favoured Areas. Proceedings of the 17th Symposium of the European Grassland Federation*, Akureyri, Iceland, 23–26 June 2013.

46. Chivers I, Warrick R, Bornman J, Evans C (2015) *Native Grasses Make New Products: A Review of Current and Past Uses and Assessment of Potential*. Publication No. 15/056. Rural Industries Research and Development Corporation, Canberra.
47. Dorrough J, Stol J, McIntyre S (2008) *Biodiversity in the Paddock: A Land Managers Guide*. Future Farm Industries CRC.
48. Pickett JA (2016) The essential need for GM crops. *Nature Plants* 2(6), 16078. doi:10.1038/nplants.2016.78
49. Australian Collaborative Rangelands Information System (2019) Department of Agriculture, Water and the Environment, Canberra. <https://www.environment.gov.au/land/rangelands/acris>
50. Parsons K (2020) Coolawanyah Station. Interviewed by Nicole Chalmer, 12 March.
51. *Buffel and Birdwood Grasses (Cenchrus ciliaris and C. setiger) in the Western Australian Rangelands*. <www.agric.wa.gov.au/rangelands/buffel-and-birdwood-grasses-cenchrus-ciliaris-and-c-setiger-western-australian-rangelands>
52. Pollock D (2019) *The Wooleen Way: Renewing an Australian Resource*. Scribe Publications, Melbourne.
53. Purvis B (2020) Woodgreen Station. Interviewed by Nicole Chalmer, March and April.
54. Vives-Peris V, de Ollas C, Gómez-Cadenas A, Pérez-Clemente RM (2020) Root exudates: from plant to rhizosphere and beyond. *Plant Cell Reports* 39(1), 3–17. doi:10.1007/s00299-019-02447-5
55. Purvis JR (1986) Nurture the land: my philosophies of pastoral management in central Australia. *Australian Rangeland Journal* 8(2), 110–117. doi:10.1071/RJ9860110
56. Australian State of the Environment (2016) *Soil*. Commonwealth of Australia 2017–2018.
57. Johnson D (2009) *Geology of Australia*. Cambridge University Press, Melbourne.
58. Thomas DT, Sanderman J, Eady SJ, Masters DG, Sanford P (2012) Whole farm net greenhouse gas abatement from establishing kikuyu-based perennial pastures in south-western Australia. *Animals (Basel)* 2, 316–330. doi:10.3390/ani2030316
59. Song Z, McGrouther K, Wang H (2016) Occurrence, turnover and carbon sequestration potential of phytoliths in terrestrial ecosystems. *Earth-Science Reviews* 158, 19–30. doi:10.1016/j.earscirev.2016.04.007
60. Schwartz JD (2013) *Cows Save the Planet: And Other Improbable Ways of Restoring Soil to Heal the Earth*. Chelsea Green Publishing, White River Junction.
61. Massy C (2017) *Call of the Reed Warbler: A New Agriculture – A New Earth*. Penguin, Sydney.
62. Dorrough J, Stol J, McIntyre S (2008) *Biodiversity in the Paddock: A Land Managers Guide*. Future Farm Industries CRC.
63. Chalmer NY. Pers. Obs.
64. Hall D (2016) Esperance Research Officer, Soils and Crop. Interviewed by Nicole Chalmer, WA Department of Agriculture and Food, 3 April.
65. Van Geel M, De Beenhouwer M, Lievens B, Honnay O (2016) Crop-specific and single-species mycorrhizal inoculation is the best approach to improve crop growth in controlled environments. *Agronomy for Sustainable Development* 36, 37. doi:10.1007/s13593-016-0373-y
66. Horowitz M (1969) Evaluation of herbicide persistence in soil. *Weed Research* 9(4), 314–321. doi:10.1111/j.1365-3180.1969.tb01490.x
67. Grady L, Graham D, Howard PH, Kannan K, Larson R et al. (2000) Monitoring as an indicator of persistance and long-range transport. In *Evaluation of Persistence and Long-range Transport of Organic Chemicals in the Environment*. (Eds G Klecka, R Boethling, J Franklin) pp. 207–212. Society of Environmental Toxicology and Chemistry, Pensacola.
68. Alavanja MCR, Hofmann JN, Lynch CF, Hines CJ, Barry KH, Barker J et al. (2014) Non-Hodgkin lymphoma risk and insecticide, fungicide and fumigant use in the agricultural health study. *PLoS ONE* 9(10), e109332.
69. Whalley RD, Chivers IH, Waters CM (2013) Revegetation with Australian native grasses: a reassessment of the importance of using local provenances. *Rangeland Journal* 35(2), 155–166. doi:10.1071/RJ12078
70. Bell LW, Bennett RG, Ryan MH, Clarke H (2011) The potential of herbaceous native Australian legumes as grain crops: a review. *Renewable Agriculture and Food Systems* 26(1), 72–91. doi:10.1017/S1742170510000347
71. Rinaudo A, Cunningham P (2008) Australian acacias as multi-purpose agro-forestry species for semi-arid regions of Africa. *Muelleria* 26(1), 79–85.
72. O'Neill G (2007) Forgotten treasures. *Ecos* 135(Feb), 8–11.
73. Roberge JM, Laudon H, Björkman C, Ranius T et al. (2016) Socio-ecological implications of modifying rotation lengths in forestry. *Ambio* 45(Suppl. 2), 109–123. doi:10.1007/s13280-015-0747-4

74. Foley T (2009) Extending forest rotation age for carbon sequestration: a cross-protocol comparison of carbon offsets of North American forests. Masters thesis, Duke University, Durham.
75. Flannery T (1999) *Throwim Away Leg: An Adventure*. Text Publishing, Melbourne.
76. Toensmeier E (2016) *The Carbon Farming Solution: A Global Toolkit of Perennial Crops and Regenerative Agriculture Practices for Climate Change Mitigation and Food Security*. Chelsea Green Publishing, New York.
77. Infante-Amate JI, de Molina MG (2013) Sustainable de-growth in agriculture and food: an agro-ecological perspective on Spain's agri-food system (year 2000). *Journal of Cleaner Production* **38**, 27–35. doi:10.1016/j.jclepro.2011.03.018
78. Sudmeyer RA, Abbott L, Jones H (2008) *Phase Farming with Trees: Field Validation of the Cropping Phase*. Publication no. 122. Rural Industries Research and Development Corporation.
79. George R, McFarlane D, Nulsen B (1997) Salinity threatens the viability of agriculture and ecosystems in Western Australia. *Hydrogeology Journal* **5**(1), 6–21. doi:10.1007/s100400050103
80. Ward B, Parks T, Blake G (2011) *Denmark River Water Resource Recovery Plan*. Report no. SLUI 40. WA Department of Water, Perth.
81. Gowdy J (2020) Our hunter-gatherer future: climate change, agriculture and uncivilization. *Futures* **115** 102488.

Appendix

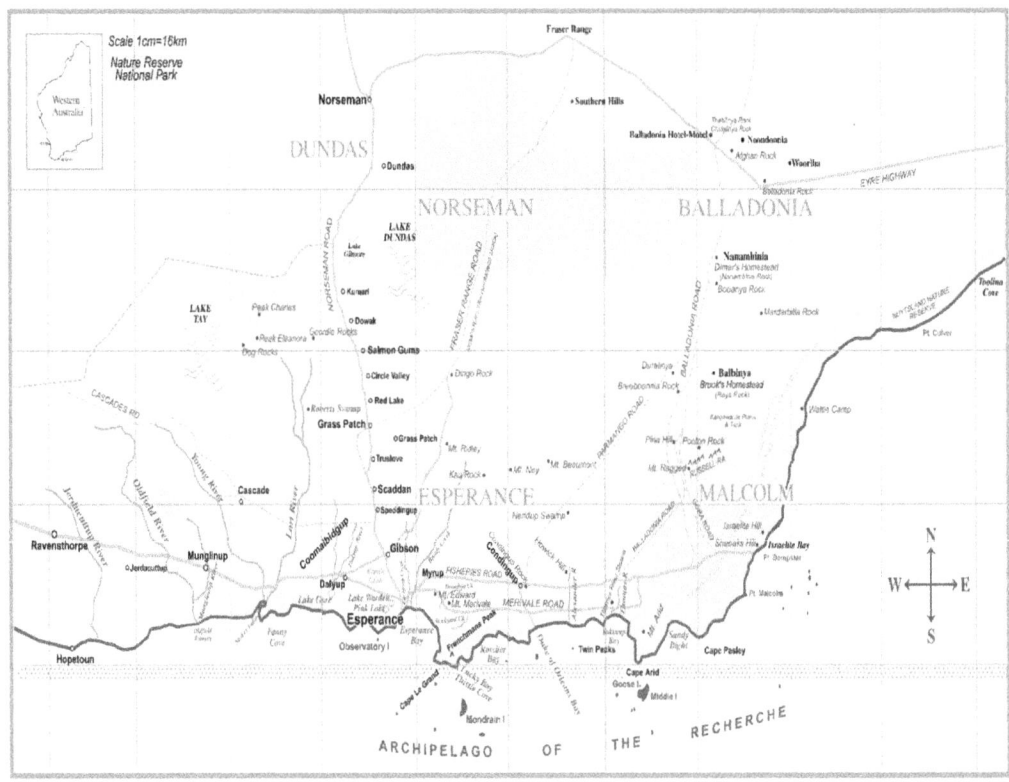

Map of the Esperance bioregion, Western Australia. Developed from numerous sources by NY Chalmer.

Index

Abbott, Ian 102
Aboriginal people 14, 33, 49, 71, 76, 78, 80, 99
   diets 33, 62–3, 68–9
   mobile shepherding 86–7, 99
   population 92, 94
   social ecological systems 71, 77, 79, 91, 93, 99, 103, 106, 117, 120, 138, 140, 145, 157
Acts 92
   Aboriginal Protection Acts, 1869–1910 92
   Dog Act 1883 92
   Environmental Protection Act 1986 132
   Game Act 1892 102
   Land Act 1933 126
   Soil and Land Conservation Act 1945 131
   Soil and Land Conservation Act 1984 131, 132
adaptation 4, 5, 40–1, 47, 148
agriculture
   commodities 124
   regenerative 156
agro-ecology 6, 154, 156
agro-ecosystem 4, 5, 137, 140
agro-systems 125, 139, 165
Albrecht, Glen 93
Americans 127
   Bitelle Institute 130
   Chase Syndicate 127–8
   Esperance Land Development Corporation (ELDC) 128
Anderson, E.N. 5, 46, 147

Barr, Neil 99
baselines, shifting 9, 10, 99
Bates, Daisy 66, 93
Berry, Wendall 137
biodiversity 2, 3, 25, 30, 124, 138, 144–5, 161
bio-pedogenesis 15, 20–1, 23, 107, 117
bioprecipitation 18, 19, 164
birds 66–7
Blaut, James 77
boosterism 114
Brandis, Tony 97
bunya pine 53
burrowing bettong 54–5

Canis dingo 9
Cape le Grand 33
carrying capacity 152
Cartesian science 152
cash crops 7
Catton, William 137, 141, 154
chemical/s 138

farming/agriculture 138, 139, 141, 143, 154, 162
circular energy cycles 8, 140
circular nutrient cycles 8, 140
clearing 115, 116, 129
climate 16, 17
   weather 17
climate change 1, 3, 5, 16, 144, 154
   economics 3
   history 8–9
   problems 3
coevolution 4, 24, 25, 39, 103, 121, 154
Colding, Johan 79
collapse 154
   agricultural 152, 154, 163
   ecosystem 30, 35, 103, 152
   population 37, 92, 152
   SES 138, 140, 153
Collins, Paul 137
colonial 137
colonisation 76–8, 109, 154
colonists 93, 94
conservation 130
consumer front 30
Country 2, 14, 41, 52, 81, 87, 94, 96, 99, 120
Crocker, Amy 92, 100, 103
CSIRO (Commonwealth Scientific and Industrial Research Organisation) 125, 126
culture 8, 23, 46–7, 51, 73, 157, 165

deficiencies
   coastal disease 125
de Molina, Manuel 123
Dempster family 83–5, 87–8, 95, 99, 108
   Dempster, Andrew 94
   Dempster, Emily 94
Denisovan 31, 32
desertification 1, 17, 114, 138, 144, 158
developmentalism 120, 130, 133
Dimer family 91, 96–8
   Dimer, Henry 96, 98, 99
   Dimer, Karl 65, 95, 97, 99
   Dimer, Tom 64, 68, 72, 94, 96
dingo 9, 40, 66, 92, 101
Diprotodon optatum 24
disease/s 91–3, 102–3, 140
DNA 32
Donahue, Brian 109
Drake, Lars 123
Dravidian 33, 40
Dreaming 14, 41, 42, 52, 73

drought   98, 99, 124, 154
dung beetles   36–7, 162

eco-environmental history   9–11
ecofarmers/farming   57, 80, 166
ecological   2, 138, 139
    knowledge   2, 46
    literacy   157–8, 159
    niche   138
    role   47
ecology   41
economic system/s   123, 140
    circular economies   152
    growth   166
    rationalism   145
ecosystem/s   1, 3, 14, 61, 62, 117, 152, 157
    collapse   101, 103
    cultural   158–9
    destruction   126
ecosystem engineers   22–3, 24, 30, 49
ecosystem management   53
EDRS (Esperance Downs Research Station)   126–7
Elders   91
Epochs   15
    Archean   15
    Eocene   15, 117
    Phanerozoic   15
    Proterozoic   15
Era
    Holocene   16, 32, 35, 40
    Pleistocene   11, 25–6, 152
Esperance bioregion   3, 16, 79, 80, 87
Europeans   73, 76
    civilisation   77, 78
    invasion   76–8
    SES   76
externality   1, 139, 142, 146
extinction   36, 38, 101, 122, 157
extractivist   78, 91, 130
Eyre, Edward John   51, 73, 76

farming
    bank debt   138, 139
    farmlands   161–3
fertiliser   1, 109, 125, 138, 139, 140, 162–3
    phosphate/phosphorous   111–12, 115, 125, 130, 138–9, 142, 162–3
finite resources   1
fire   22, 39, 54–6, 71, 87, 99
    Aboriginal burning practices   100, 103
    control   3, 55, 144
    habitat   56
    karl   55
    land clearing   46
fish   67
Flannery, Tim   1, 100

Folke, Carl   79
food   1, 16, 139
    commodity   120, 122
    culture   5–7
    diversity   62
    nutritional benefits   62, 147
    plants and fruit   69
    production   1, 77, 153
    pyramid   61
    ration/rationing   92, 93
    security   138, 147
    self-sufficiency   91, 114, 122
    storage   70
    toxins   69, 70
    traditional food   6, 91, 93
Forrest, Sir John   91, 93, 107, 110

genetic modification   154
Genyornis sp.   25, 33
ghost resources/economies   137, 141, 142
Gilbert, John   26
Gilmore, Dame Mary   47, 52
globalisation   146, 153
gnamma   71–2
Gott, Beth   68
Goyder, George   124
granite domes   16, 71
grasses   24, 108, 164
    coevolution   25
    C3 and C4 plants   34, 36
Great Depression   117
Green, Neville   95
Green Revolution   138
Grewar, Geoff   128
Grey, George   62, 63, 66, 69, 106
Gunn, Sir William   127

habitat   2, 3, 22, 42, 64, 99, 100, 132
    loss   3, 102, 117
Hallam, Sylvia   55
Hammond, Jesse E.   94
Hannett, Edward   65
Hassell, Ethel   62, 71
herbivore   61, 64, 81
Holling, Carl   137
Homo erectus   5
Homo sapiens   5, 25, 30, 32
human hunting   34
hunter gatherers   46, 48–9, 61, 77, 166
Hunza people   50
hydrology   18, 19, 99, 117, 126, 138, 158

Incan civilisation   153
industrialised agriculture   1, 5, 77, 106, 109–10, 122–4, 137, 138, 140, 148, 152, 154, 157, 166
Infante-Amate, Juan   123

influenza   91
insects   67–8
invasion   11, 77, 94, 140
    culture   137

keystone   139
    culture   26, 41, 52, 103
    mammal/s   103
    predator   101
    role   26
    species   23, 30
King, Professor Franklin J.   140, 152

Landcare   131–2
landscape/s   14–15
    animal   25, 103
    change   24, 56, 106
    culture   22
    degradation   1, 97
    first class   107–8
    freehold   107
    management   24, 56
    nature   4, 22
linear energy flow   141
lizard traps   56

Madagascar   26
maladaptive   46, 141, 152, 153
Malcolm, C.V.   132
mallee   3, 16, 107, 113, 115, 122, 127
mammals   65
markets   156
    free   145–6
Markey, D.C.   109
marsupials, Australian   78, 93, 100–2
Martin, Greg   103
massacre/s   95
measles   91–3
megafauna   25, 33, 35–9, 81
Meinig, Donald   109, 124
Melville, Elinor   76, 80
Microseris scapigera/lanceolata   see yam daisy
monoculture/s   123, 138
mythology   14

Nanambinia Station   96, 98, 99
nardoo   69
nature   137, 139, 166
Neanderthal   32
neoliberalism   145, 146, 156, 157
Ngadju   51, 62–3, 73, 76, 106
niche   22, 39, 48
Nind, Isaac   65
Nyungar   51, 67, 73, 76, 93, 95–6, 106

omnivore   61

overgrazing   100
overkill   46
overstocking   97, 98, 139, 140

pastoralism   9, 76, 83–4, 87–8, 92–3, 95, 99, 106–7, 117
    mobile   96, 111
    pastoralists   76, 91
    template   76, 83
Penn, Justin J.   46, 47
phosphate/phosphorous   11, 112, 115, 125, 130, 138–40, 142, 162, 163
Pickard, John   95
planetary boundaries   155
plants   68
    annual   138
    improved pasture   25, 126, 128
    native pastures   125
    perennial   2, 3, 98, 129, 131–2, 138, 139, 158, 161, 163–4
    poison (*Gastrolobium* sp.)   70, 129
    toxins   37
population
    collapse   73
    control   49–50
    growth   77, 110, 154, 156–7, 165–6
possums   102, 103
predator/s   61, 101, 102
    release   101
Pretty, Jules   6, 154, 157
Prober, Susan   79
Purdie, Brian   127

rabbits   91, 93, 101
rainfall   17
    biotic pump   34
    decline   17, 18
    modelling   17, 18, 33
rangelands   9, 97, 159–61
Ratcliffe, Francis   97
reptiles   66, 67
resilience   2, 7–8, 138, 139, 141, 152, 155, 157, 166
Resilience Alliance   139
resource exploitation   46, 47
Rockström, Johan   155
Roe, John Septimus   49, 55, 79, 83–4
Rose, Deborah Bird   52
rotational grazing   81, 83
Royal Commissions
    1905   115
    1916   115
    1938   126
ruminant   85

Sahul   30–3, 73
Saifi, Basim   123

salinity 2, 3, 19–20, 113, 115, 117, 120, 126, 130–2, 142–3, 158, 165
sanctuaries 52
sandplain 3, 111, 113, 122, 124, 127, 130, 162
seasonal calendars 51
SES *see* social ecological systems
set stocking 81
sheep 80–1, 84–5, 91, 95, 98, 100–2, 106
shepherding 97, 98
Shiva, Vandana 138
Smith, Moya 56, 64–5
social ecological systems (SES) 4–5, 48, 76, 96, 108–9, 139, 141, 145, 152, 155, 157
soils 1, 2, 103
    acidity 143
    biology 1, 2, 21–2
    classification 118, 126, 141
    degradation 2, 97, 109, 110, 124, 139
    fertility 2, 15, 109–11, 124, 141, 153
    mapping 107, 127
    micro-nutrients/nutrients 125, 128, 141, 152, 163
    re-engineer 2
    science 114, 115
    topsoils (A and B horizon) 21, 140, 152, 153
solastalgia 11, 93
Spanish flu 94
spatial-visual memory 14
Sporormiella fungus 34

stage theory 77
starvation 92–3
Stephens, Danielle 101
supermarkets 147
sustainable/sustainability 7, 78, 139, 140, 141, 145, 147, 156, 166
    agriculture 152, 155, 156, 157
    development 7, 138

taboos 46, 51–2
Tasmanian devil 25, 40, 41
Teakle, L.J.H. 116–18, 126
terra nullius 11, 77
transhumance 11, 76, 79, 84–6, 88, 95
trophic cascade 35–7

ungulate eruption/s 80, 85, 88
unsustainable 5

Ward, David 55–6
water 71, 96, 97
water trees 72
    kumbal, pillirri 72
Wheelman group 52, 71
white-footed rabbit rat 68
Wurundjeri 68

yam daisy 68
yeoman farms 88, 91, 106, 108–10, 114, 125

www.ingramcontent.com/pod-product-compliance
Lightning Source LLC
Chambersburg PA
CBHW040356010526
44108CB00049B/2923